BIOCHEMISTRY RESEARCH TRENDS

ISOTOPES FOR THE NATION'S FUTURE: RESEARCH OPPORTUNITIES AND PLANS

BIOCHEMISTRY RESEARCH TRENDS

Additional books in this series can be found on Nova's website
under the Series tab.

Additional E-books in this series can be found on Nova's website
under the E-books tab.

BIOCHEMISTRY RESEARCH TRENDS

ISOTOPES FOR THE NATION'S FUTURE: RESEARCH OPPORTUNITIES AND PLANS

EZIO BENFANTE
EDITOR

Nova Science Publishers, Inc.
New York

Copyright © 2012 by Nova Science Publishers, Inc.

All rights reserved. No part of this book may be reproduced, stored in a retrieval system or transmitted in any form or by any means: electronic, electrostatic, magnetic, tape, mechanical photocopying, recording or otherwise without the written permission of the Publisher.

For permission to use material from this book please contact us:
Telephone 631-231-7269; Fax 631-231-8175
Web Site: http://www.novapublishers.com

NOTICE TO THE READER

The Publisher has taken reasonable care in the preparation of this book, but makes no expressed or implied warranty of any kind and assumes no responsibility for any errors or omissions. No liability is assumed for incidental or consequential damages in connection with or arising out of information contained in this book. The Publisher shall not be liable for any special, consequential, or exemplary damages resulting, in whole or in part, from the readers' use of, or reliance upon, this material. Any parts of this book based on government reports are so indicated and copyright is claimed for those parts to the extent applicable to compilations of such works.

Independent verification should be sought for any data, advice or recommendations contained in this book. In addition, no responsibility is assumed by the publisher for any injury and/or damage to persons or property arising from any methods, products, instructions, ideas or otherwise contained in this publication.

This publication is designed to provide accurate and authoritative information with regard to the subject matter covered herein. It is sold with the clear understanding that the Publisher is not engaged in rendering legal or any other professional services. If legal or any other expert assistance is required, the services of a competent person should be sought. FROM A DECLARATION OF PARTICIPANTS JOINTLY ADOPTED BY A COMMITTEE OF THE AMERICAN BAR ASSOCIATION AND A COMMITTEE OF PUBLISHERS.

Additional color graphics may be available in the e-book version of this book.

Library of Congress Cataloging-in-Publication Data

ISBN 978-1-61470-818-6

Published by Nova Science Publishers, Inc. ✛ *New York*

CONTENTS

Preface		**vii**
Chapter 1	Compelling Research Opportunities Using Isotopes *Nuclear Science Advisory Committee Isotopes Subcommittee*	**1**
Chapter 2	Isotopes for the Nation's Future: A Long Range Plan *Nuclear Science Advisory Committee Isotopes Subcommittee*	**43**
Index		**175**

PREFACE

This book explores isotopes which are vital to the science and technology base of the U.S. economy. Isotopes, both stable and radioactive, are essential tools in the growing science, technology, engineering and health enterprises of the 21st century. The scientific discoveries and associated advances made as a result of the availability of isotopes today span widely from medicine and biology, physics, chemistry and a broad range of applications in environmental and material sciences. Isotope issues have become crucial aspects of homeland security. Isotopes are utilized in new resource development, in energy from bio-fuels, petrochemical and nuclear fuels, in drug discovery, health care therapies and diagnostics, in nutrition, in agriculture and in many other areas.

Chapter 1- Isotopes are vital to the science and technology base of the US economy. Isotopes, both stable and radioactive, are essential tools in the growing science, technology, engineering, and health enterprises of the 21st century. The scientific discoveries and associated advances made as a result of the availability of isotopes today span widely from medicine to biology, physics, chemistry, and a broad range of applications in environmental and material sciences. Isotope issues have become crucial aspects of homeland security. Isotopes are utilized in new resource development, in energy from bio-fuels, petrochemical and nuclear fuels, in drug discovery, health care therapies and diagnostics, in nutrition, in agriculture, and in many other areas.

The development and production of isotope products unavailable or difficult to get commercially have been most recently the responsibility of the Department of Energy's Nuclear Energy program. The President's FY09 Budget request proposed the transfer of the Isotope Production program to the Department of Energy's Office of Science in Nuclear Physics and to rename it the National Isotope Production and Application program (NIPA). The transfer has now taken place with the signing of the 2009 appropriations bill. In preparation for this, the Nuclear Science Advisory Committee (NSAC) was requested to establish a standing subcommittee, the NSAC Isotope Subcommittee (NSACI), to advise the DOE Office of Nuclear Physics. The request came in the form of two charges: one, on setting research priorities in the short term for the most compelling opportunities from the vast array of disciplines that develop and use isotopes and two, on making a long term strategic plan for the NIPA program. This is the final report to address charge 1.

Chapter 2- In 2009, with the signing of the FY09 Omnibus Spending Bill (Public Law 111-8), the Department of Energy's Isotope Production Program was transferred from the Department of Energy (DOE) Office of Nuclear Energy (NE) to the Office of Science's Office of Nuclear Physics (ONP). The name of the program has been changed from the

National Isotopes Production and Applications Program (NIPA) to the Isotope Development and Production for Research and Applications Program (IDPRA). To prepare for this transfer, the Office of Nuclear Physics and the Office of Nuclear Energy organized a workshop held August 5-7, 2008, in Rockville, MD, that brought together the varied stakeholders in the isotopes enterprise to discuss "the Nation's current and future needs for stable and radioactive isotopes, and options for improving the availability of needed isotopes." The report [NO08] of the "Workshop on the Nation's Needs for Isotopes: Present and Future" is available on the web http://www.sc.doe.gov/henp/np/program/docs/Workshop%20Report_final.pdf). On August 8, 2008, the DOE-ONP requested the Nuclear Science Advisory Committee (NSAC) to establish a standing committee, the NSAC Isotope (NSACI) subcommittee, to advise the DOE Office of Nuclear Physics on specific questions concerning the isotope program. NSAC received two charges from the DOE Office of Nuclear Physics. The first charge requested NSACI to identify and prioritize the compelling research opportunities using isotopes. NSAC accepted the final report on the first charge in April 2009 and transmitted the report [NS09] to the Department of Energy (http://www. sc.doe.gov/henp/np/nsac/docs/NSAC_ Final_ Report_Charge 1 %20(3).pdf). The second charge is to study the opportunities and priorities for ensuring a robust national program in isotope production and development, and to recommend a long-term strategic plan that will provide a framework for a coordinated implementation of the Isotope Development and Production for Research and Applications Program.

In: Isotopes for the Nation's Future
Editor: Ezio Benfante

ISBN: 978-1-61470-818-6
© 2012 Nova Science Publishers, Inc.

Chapter 1

COMPELLING RESEARCH OPPORTUNITIES USING ISOTOPES[*]

Nuclear Science Advisory Committee Isotopes Subcommittee

Final Report
One of Two 2008 Charges to NSAC on the National
Isotopes Production and Applications Program

The Cover: The discovery of isotopes is less than 100 years old, today we are aware of about 250 stable isotopes of the 90 naturally occurring elements. The number of natural and artificial radioactive isotopes exceeds 3200, already, and this number keeps growing every year. "Isotope" originally meant elements that are chemically identical and non-separable by chemical methods. Now isotopes can be separated by a number of methods such as distillation or electromagnetic separation.The strong colors and the small deviations from one to the other indicate the small differences between isotopes that yield their completely different properties in therapy, in nuclear science, and a broad range of other applications. The surrounding red, white and blue theme highlights the broad national impact of the US National Isotope Production and Application program.

EXECUTIVE SUMMARY

Isotopes are vital to the science and technology base of the US economy. Isotopes, both stable and radioactive, are essential tools in the growing science, technology, engineering, and health enterprises of the 21st century. The scientific discoveries and associated advances made as a result of the availability of isotopes today span widely from medicine to biology, physics, chemistry, and a broad range of applications in environmental and material sciences. Isotope issues have become crucial aspects of homeland security. Isotopes are utilized in new

---

[*] This is an edited, reformatted and augmented version of a Nuclear Science Advisory Committee Isotopes Subcommittee publication, dated April 23, 2009.

resource development, in energy from bio-fuels, petrochemical and nuclear fuels, in drug discovery, health care therapies and diagnostics, in nutrition, in agriculture, and in many other areas.

The development and production of isotope products unavailable or difficult to get commercially have been most recently the responsibility of the Department of Energy's Nuclear Energy program. The President's FY09 Budget request proposed the transfer of the Isotope Production program to the Department of Energy's Office of Science in Nuclear Physics and to rename it the National Isotope Production and Application program (NIPA). The transfer has now taken place with the signing of the 2009 appropriations bill. In preparation for this, the Nuclear Science Advisory Committee (NSAC) was requested to establish a standing subcommittee, the NSAC Isotope Subcommittee (NSACI), to advise the DOE Office of Nuclear Physics. The request came in the form of two charges: one, on setting research priorities in the short term for the most compelling opportunities from the vast array of disciplines that develop and use isotopes and two, on making a long term strategic plan for the NIPA program. This is the final report to address charge 1.

NSACI membership is comprised of experts from the diverse research communities, industry, production, and homeland security. NSACI discussed research opportunities divided into three areas: 1) medicine, pharmaceuticals, and biology, 2) physical sciences and engineering, and 3) national security and other applications. In each area, compelling research opportunities were considered and the subcommittee as a whole determined the final priorities for research opportunities as the foundations for the recommendations. While it was challenging to prioritize across disciplines, our order of recommendations reflect the compelling research prioritization along with consideration of time urgency for action as well as various geo- political market issues. Common observations to all areas of research include the needs for domestic availability of crucial stable and radioactive isotopes and the education of the skilled workforce that will develop new advances using isotopes in the future. The six recommendations of NSACI reflect these concerns and the compelling research opportunities for potential new discoveries. The science case for each of the recommendations is elaborated in the respective chapters.

The six recommendations of NSACI are summarized below in order of priority:

There are compelling research opportunities using alpha-emitters in medicine. There is tremendous potential in developing far more effective treatments of cancers by the use of alpha-emitters in comparison to other radio-isotopes. Therefore, development and testing of therapies using alpha emitters are our highest priority for research isotope production for the medical field. This opportunity can be realized by a variety of alpha emitters with the highest priority given to ^{225}Ac. This priority is reinforced by the potential need for rapid action due to the 2012 deadline for downblending of current DOE stocks of ^{233}U, a procedure that would eliminate its value as a source of ^{225}Ac.

1. Invest in new production approaches of alpha-emitters with highest priority for 225Ac. Extraction of the thorium parent from 233U is an interim solution that needs to be seriously considered for the short term until other production capacity can become available.

There is strong evidence for the potential efficacy of pairs of isotopes with simultaneous diagnostic/therapeutic capabilities where damage to normal tissue is minimized and exposure to target tissue is enhanced. NSACI finds the research opportunities offered with these pairs of isotopes to be the second highest priority. Many of the short-lived diagnostic/therapeutic isotopes (2nd entry in Table 8) could be produced at existing accelerator facilities but are not widely available. We recommend the maximization of the production and availability of these isotopes domestically in the US through investments in research and coordination between existing accelerators. Increasing such coordination and R&D has the potential to improve the availability of a number of other isotopes. The panel felt that such a network could benefit all areas of basic research and applications from security to industry. This should include R&D to standardize efficient production target technology and chemistry procedures.

2. We recommend investment in coordination of production capabilities and supporting research to facilitate networking among existing accelerators.

The basic physical sciences and engineering group prioritized research opportunities across various disciplines and a summary of this prioritization is given in Table 9. The availability of californium, radium, and other trans-uranic isotopes are particularly important for research in nuclear physics and chemistry.

3. We recommend the creation of a plan and investment in production to meet these research needs for heavy elements.

Experts in the nuclear security and applications areas strongly consider the vulnerability of supply from foreign sources to be of highest priority. This concern was echoed strongly by all members of the subcommittee from medicine to basic science and engineering. Additionally, the projected demand for ^{3}He by national security agencies far outstrips the supply. This would likely endanger supply for many other areas of basic research. While it is beyond our charge, it would be prudent for DOE/NNSA and DHS to seriously consider alternative materials or technologies for their neutron detectors to prepare if substantial increases in ^{3}He production capacity cannot be realized.

4. We recommend a focused study and R&D to address new or increased production of ^{3}He.

An important issue for the use of isotopes is the availability of high-purity, mass separated isotopes such as ^{236}Np for dilution mass spectrometry. The stable isotopes ^{76}Ge and ^{28}Si (^{3}He is stable but obtained from the beta-decay of ^{3}H, not by isotope separation) listed in Table 9 are needed in large quantities that present special problems. While no other individual stable isotope reached the level of the highest research priority, the broad needs for a wide range of mass-separated isotopes and the prospect of no domestic supply raised this issue in priority for the subcommittee. NSACI feels that the unavailability of a domestic supply poses a danger to the health of the national research program and to national security. The subcommittee recommends:

5. Research and Development efforts should be conducted to prepare for the reestablishment of a domestic source of mass-separated stable and radioactive research isotopes.

Vital to the success of all scientific endeavors is the availability of a trained workforce. While the scientific opportunities have expanded far beyond the disciplines of radiochemistry and nuclear chemistry, the availability of trained personnel remains critical to the success of research in all frontiers of basic science, homeland security, medicine, and industry. The individual research areas must make concerted efforts to invest in work-force development to meet these needs. The isotope program has a special responsibility to ensure a trained workforce in the production, purification and distribution of isotopes.

6. We recommend that a robust investment be made into the education and training of personnel with expertise to develop new methods in the production, purification, and distribution of stable and radio-active isotopes.

All of the issues and recommendations considered here will be important input for answering the 2nd NSACI charge in developing a long range plan for the Nuclear Isotopes Production and Application Program.

1. INTRODUCTION

The Fiscal Year (FY) 2009 President's Request Budget proposed to transfer the Isotope Production Program from the Department of Energy (DOE) Office of Nuclear Energy to the Office of Science's Office of Nuclear Physics, and rename it the National Isotope Production and Applications Program (NIPA). The transfer is in effect since the signing of the 2009 appropriations bill and there is a new name for the program, mainly, the Isotope Development and Production for Research and Applications. For the sake of consistency with the NSAC charge, we use NIPA for the remainder of this report.

In preparation for this transfer from NE to NP, the Nuclear Science Advisory Committee (NSAC) was requested to establish a standing subcommittee, the NSAC Isotope (NSACI) subcommittee, to advise the DOE Office of Nuclear Physics on specific questions concerning the National Isotope Production and Applications (NIPA) Program. NSAC received two charges from the DOE Office of Nuclear Physics. A copy of the full charge letter is attached as Appendix 1.

Charge 1

As part of the NIPA Program, the FY 2009 President's Request includes $3,090,000 for the technical development and production of critical isotopes needed by the broad U.S. community for research purposes. NSACI is requested to consider broad community input regarding how research isotopes are used and to identify compelling research opportunities using isotopes. The subcommittee's response to this charge should include the identification and prioritization of the research opportunities; identification of the stable and radioactive isotopes that are needed to realize these opportunities, including estimated quantity and purity; technical options for producing each isotope; and the research and development efforts associated with the production of the isotope. Timely recommendations from NSACI will be important in order to initiate this program in FY 2009; for this reason an interim report is requested by January 31, 2009, and a final report by April 1, 2009.

This document is the final report for Charge 1.

Background

The Atoms for Peace program launched by President Eisenhower with an address to the United Nations National Assembly in 1953 and then followed by the Atomic Energy Act of 1954 has grown and translated into tremendous global advances in the 21st century. The infrastructure developed for making nuclear weapons was converted to the production of isotopes and made available to the research infrastructure of the world. The scientific discoveries and advances made as a result of the availability of both stable and radioactive isotopes span widely from medicine to biology, physics, chemistry, and a broad range of applications in environmental and material sciences. Isotopes have also become crucial aspects of homeland security. They are utilized in new resource development, in energy from bio-fuels, in nuclear and petrochemical, drug discovery, health care therapies and diagnostics, nutrition, agriculture, and many others. In effect, the Atoms for Peace program led to the development of the modern field of nuclear medicine. The instrumentation of the gamma camera and Positron Emission Tomography (PET), as well as the most important radiopharmaceuticals, such $^{99m}Tc/^{99}Mo$ generator systems and many ^{99m}Tc radiopharmaceuticals including ^{99m}Tc sestamibi (a pharmaceutical agent used in nuclear medicine) for cardiac perfusion, ^{18}F flourodeoxyglucose (FDG), as well as ^{32}P, ^{198}Au, ^{131}I for therapy, were all developed with sole or partial support by DOE.

Isotopes were produced by the US Department of Energy for more than fifty years. In FY 2009, the President's Budget request included a proposal to move the Isotope Program currently in the Office of Nuclear Energy (NE) to the Office of Nuclear Physics (NP) within the Department's Office of Science. The goal of the Office of Nuclear Physics is to manage the Isotopes Program in an optimum fashion for the development and production of key isotopes for use in all the forefront areas of research in the sciences, in medicine, in industry, and national security. While the production of isotopes is centralized in the federal government, the variety of research supported by the use of DOE-produced isotopes span many agencies as varied as DOE

Office of Science, NSF, National Institutes of Health, NNSA, EPA and NIST amongst others. There are also numerous industrial interests that use and produce isotopes for a variety of homeland security, research, and medical interests.

The office of Nuclear Physics organized a workshop August 5-7, 2008 in Rockville, MD bringing together all the varied stakeholders in the isotopes enterprise to discuss the development of a robust and prioritized program for isotope production. The report[1] of the "Workshop on the Nation's Needs for Isotopes: Present and Future" is now available on the web (http://www.sc.doe.gov/henp/np/program/docs/Workshop%20Report_final.pdf). The workshop identified key isotopes from all the areas of research that were in short supply and highlighted several issues that are crucial to the future of the isotopes program.

- A reliable program in isotope production at DOE is crucial for the long term health of developments in medicine, basic physical and biological sciences, national security and industry.
- Many isotopes in domestic use are presently only produced by foreign suppliers, often a single or limited number of suppliers. This makes the isotope supply

vulnerable to interruption or large price fluctuations beyond the control of the United States.

- Affordability is an important issue for research isotopes.
- The production capability of the NIPA program relies on facilities that are operated by DOE for other primary missions.
- There is a pressing need for more training and education programs in nuclear science and radiochemistry to provide the highly skilled work force for isotope application.
- The DOE Isotope Program and the resources that it has available to it today cannot fulfill the broad challenges and needs for current and future demands of the nation for isotopes.

The workshop by design did not address the relative priorities for uses for various isotopes. Setting priorities *between various disciplines and end users is clearly another major issue.*

Procedures

The NSACI subcommittee membership was chosen to have broad representation from the research communities, industry, production, and homeland security. The list of NSACI members is given in Appendix 2. A special effort was made for the membership to have overlap with the many ongoing studies and the August Workshop. The studies of interest included the National Academies of Sciences (Institute of Medicine) study "*Isotopes for Medicine and Life Sciences*" published in 1995, the report on the 2008 meeting to discuss "*Existing and Future Radionuclide Requirements of the National Cancer Institute*", and the studies of the National Academies: "*Advancing Nuclear Medicine through Innovation*" published in2007 and "*Medical Isotope Production without Highly Enriched Uranium*" just published in 2009.

Three meetings were called by the subcommittee to address the first charge. The agendas for the meetings are attached in Appendix 3. The goal was to determine the landscape in isotope needs/uses/production, to determine needs for research isotopes across agencies, industry, and professional societies on the road to prioritization. The first meeting (Nov 13-14, 2008) was dedicated to hearing results of the several of the recent studies including the DOE workshop. The second meeting (Dec 15-16, 2008) was dedicated to hearing from the Office of Management and Budget and the needs of the various federal agencies that fund research with isotopes. The third meeting (January 13-15, 2009) was dedicated to hearing about research needs from a wide range of professional societies and some of the facilities. Appendices 4 and 5 contain a listing of all agencies and professional societies that were contacted.

Overall, there is broad enthusiasm in favor of the Office of NP management and the potential opportunities presented to the isotope program. There is also an overwhelming concern for the education and training of students in radiochemistry, nuclear chemistry, and nuclear physics with the expertise required to make significant contributions toward the development of new and more traditional techniques using radioisotopes.

The NSACI committee divided the discussion of research opportunities into three areas: 1) medicine, pharmaceuticals, and biology, 2) physical sciences and engineering, and 3) national security and other applications. In each area, compelling research opportunities were considered and the subcommittee as a whole determined the priority research opportunities and recommendations. In order to realize the opportunities presented in each of these areas, isotope production capabilities will be of paramount importance. In the next chapter, we examine the present landscape for isotope production in the US.

2. LANDSCAPE OF ISOTOPE PRODUCTION IN NORTH AMERICA AS OF 2009

The landscape of isotope production in the US spans across many horizons and includes government, private industry, and research facilities at Universities. This chapter is divided into the following sections:

- Stable Isotopes
- Radionuclides produced in reactors
- Accelerator produced radionuclides which are further divided into sections reflecting the commercial suppliers, DOE labs and University/Hospital based producers.

By the nature of this enterprise there are, in many cases, overlaps of production sectors in supplying a particular isotope or radionuclide.

Stable Isotopes (Private and Government Operated)

Active production of stable isotopes in the United States is primarily performed by the private sector with the exception of Helium-3 which is produced at the Savannah River facility operated by the DOE as a by-product of tritium decay. For the purposes of this report *production* refers to a process whereby a stable isotope of an element is both separated and enriched to a useable level which is typically above 90 atom%.

The methods of isotope separation and enrichment employed by the private sector companies are distillation, chemical exchange, and thermal diffusion. The private companies which have these capabilities are Cambridge Isotope Laboratories, Eagle Pitcher, Isotec (Sigma Aldrich), and Spectra Gases. They produce the isotopes of carbon (13), oxygen (17, 18), and boron (10, 11) with capacity in the metric ton range and they offer a wide variety of compounds labeled with these isotopes. Supply is not an issue for any of these particular isotopes. Additionally, Isotec has a set of thermal diffusion columns which can be used for production of gaseous isotopes. These systems are not competitive in cost of production to those using the cryogenic centrifuges employed by foreign manufactures but do provide some domestic capability for some isotopes such as krypton, and xenon. Additionally, Spectra Gases has systems capable of enriching nitrogen (15), but the demand for this isotope is currently met by foreign entities at extremely reasonable prices.

Plasma isotope separation is a tool that has been used successfully in the United States weapons programs and is capable of producing specific isotopes at medium enrichments. These systems have been dismantled and are no longer available; however the technology is useful to an enrichment program if integrated with electromagnetic separators. A private company, Nonlinear Ion Dynamics (NID), LLC, in California has a plasma isotope enrichment system and has operated under an SBIR grant and private funding.

Research in the United States that uses stable and enriched isotopes is strategically important and in the Nation's interest. Therefore, the domestic capability for the production of these stable isotopes is strategically important to the United States in order to assure the continuation of research activities.

There are 339 naturally occurring isotopes (nuclides) on earth, and 250 of these are stable isotopes. The vast majority of the 250 naturally occurring stable isotopes are primarily made up of alkalis, alkali-earth, and metal stable isotopes and require the use of electromagnetic separators which are no longer in use within the United States.

The 220 stable non-gaseous isotopes are not currently produced in the United States. The reasons for this are several but include:

1) Most are only used in research applications in very limited quantities which will not attract private sector investment to manufacture them.
2) Most require separation and enrichment by means of either electromagnetic or gas centrifuge separators, and these systems are not operational in the United States and are prohibitively expensive to make and may require classified technology to build them and DOE security systems to manage them.
3) Many of the isotopes produced by electromagnetic separators are currently inventoried at ORNL in sufficient quantities to support limited research, however a number of isotopes (see Table 1) are no longer in inventory and/or are well below levels to sustain research even in the short term.
4) Foreign supply of the stable isotopes requiring centrifuges is currently meeting demand in most cases. However in a few instances the foreign supply is not meeting demand, e.g. ^{136}Xe, and ^{76}Ge.
5) The thermal diffusion method used for the separation and enrichment of the rare gas isotopes of argon, neon, krypton, and xenon is an expensive method of production.
6) Plasma isotope separators require more research to be used as a production tool.
7) Laser Isotope Separators appear prohibitively expensive to operate and currently limited in their scope of production for stable isotopes.

The alkalis, alkali-earth, and metal stable isotopes are essential to current research in health care and nutrition studies, which manipulate biochemistry at the cellular and sub-cellular level to prevent disease, as well as to offer personalized detection and treatment. The research cannot be done without the assurance of an ongoing supply of these isotopes. Some of these, identified in Table 1 from the August workshop, are already in short supply or no longer available and the only demonstrated method of production and separation for these stable isotopes is electromagnetic separators like the original Calutrons currently on standby at Oak Ridge National Laboratory. These systems have not operated due to the high cost of operations at ORNL and the 1990's implementation of the policy to have full cost recovery for all production at this facility. The income from the sale of the existing stockpiles of isotopes has not been sufficient to fund operation of the separators.

An important issue for stable isotope production is the potential use of these technologies for weapons of mass destruction. As such, much of the forefront work in this area may be classified and the technology is subject to security and export controls.

Table 1. Stable Isotopes of limited supply identified at the August Workshop[1]

Isotope	Years remaining in inventory
^{157}Gd, Second Pass	0
^{204}Pb, Second Pass`	0
^{207}Pb, Second Pass	0
^{96}Ru	0
^{150}Sm, Second Pass	0
^{181}Ta	0
^{180}W, Second Pass	0
^{51}V	0
^{157}Gd	0.2
^{154}Gd, Second Pass	2.5
^{69}Ga	3.7
^{62}Ni	3.9

Reactor Production of Radionuclides

Worldwide, 278 research reactors are known to be operating in 56 countries supporting a variety of test, training, and research missions including isotope production. Eight new research reactors are under construction and eight more are currently being planned, all of them outside the U.S. The majority of these reactors are over 30 years old and the number of shutdown or decommissioned research reactors is about 487, by far more than are operating today. In the United States, there are currently 32 operating research reactors that are licensed by the U.S. Nuclear Regulatory Commission (NRC) and two reactors operated by the U.S. Department of Energy (DOE) for research and isotope production. Of these reactors, twenty-seven are owned and operated by universities and colleges for education and research purposes. Most of the U.S. research reactors are over 40 years old; however, many have recently completed or are currently in the process of re-licensing for an additional 20 years. A listing of U.S. research reactors having a thermal power of 1 MW or more is included in Table 2.

The number of DOE research reactors has decreased since 1980, primarily due to aging infrastructure and a change in programmatic focus by the DOE in the 1980s. Likewise, the number of university research reactors (URR) in the U.S. has decreased by over 50% since 1980. This trend was partially due to the closure of many nuclear engineering programs following a contraction of the nuclear industry, but was also caused by the lack of funding necessary to replace failed or obsolete research and reactor control instrumentation. Nearly all

URR facilities had, and still have, very limited funding to upgrade existing systems. The age of the URR instrumentation and control systems impacted their reliability and availability and thus limited the number of users. In 1998, the DOE Office of Nuclear Energy (DOE-NE) acted to support the URR facilities and created the University Reactor Instrumentation grants to replace reactor control and safety system equipment at the aging URR facilities. This program and other funding programs provided targeted support to the URR community and appeared to halt the trend of URR closures. Unfortunately, these programs were cancelled in 2006. There are no current plans for replacement funding programs, which again places many URR facilities at risk of closure.

Table 2. U.S. Research Reactors (> 1 MWth)

Research Reactor Facility	Power Level (MW)
Advanced Test Reactor (ATR)	250
High Flux Isotope Reactor (HFIR)	85
National Institute for Standards and Technology-NBSR	20
University of Missouri-Columbia	10
Massachusetts Institute of Technology	5
University of California-Davis	2.3
Rhode Island Nuclear Science Center	2
Washington State University	1.3
Kansas State University	1.25
Oregon State University	1.1
Penn State University	1.1
University of Texas at Austin	1.1
North Carolina State University	1
Texas A&M University	1
University of Massachusetts-Lowell	1
University of Wisconsin-Madison	1
U.S. Geological Survey	1
Armed Forces Radiobiology Research Institute	1

Isotope Production in U.S. Research Reactors

The University of Missouri reactor (MURR) may be considered unique among U.S. research reactors because a large percentage of its operations are solely dedicated to the large scale production of medical radioisotopes. MURR operates on a regular 6.5 day/week operating schedule and has availability of approximately 90% during its scheduled operating time. This schedule and high availability make it ideal for routine, reliable production and shipment of radioisotopes. In 2007, MURR produced and shipped over 30,000 Ci of ^{153}Sm, ^{90}Y, ^{33}P, ^{32}P, ^{166}Ho, and ^{177}Lu for medical research, clinical trials, and commercial applications. MURR is also actively engaged with industry to become a reliable domestic source of ^{99}Mo for the production of ^{99m}Tc generators used in approximately 85% of all nuclear medicine procedures. ^{99}Mo is currently imported from Canada and Europe.

The smaller research reactors also provide important radioisotopes to researchers and commercial radiotracer companies. These users have a need for small batch quantities of relatively short-lived radioisotopes that would not be obtainable or useful if provided by a reactor facility located hundreds of miles away. Examples of some of these short-live isotopes (<100 days) are provided in Table 3.

These radioisotopes do not have a substantial "shelf-life" and require just-in-time production and delivery. The smaller research reactors have flexible staff, facilities and operational schedules to provide this capability. The distribution of these small research reactors throughout the U.S. ensures that customers in all regions can be supplied with the short-lived radioisotopes to support their needs.

Use of the DOE research reactors for the production of radioisotopes is somewhat limited by a policy adopted in 1965 that the U.S. government is to refrain from competing with private sources when the materials are reasonably available commercially. However, these reactors may and do engage in the production of isotopes that can only be produced in the high flux available in their cores or that are being developed for future commercialization. At a consequence of their multiple missions, in some cases, the primary mission at the reactor leads to operating schedules that are not well matched for optimum isotope production. Examples of radioisotopes that can be produced at the DOE reactors are provided in Table 4.

It should be pointed out that some of these radionuclides that are produced via the (n,γ) reaction will have limited utility because of the low specific activity associated with this approach to production (e.g. ^{186}Re, ^{64}Cu).

The heavy actinides, such as ^{252}Cf, can be produced only at HFIR. This capability is unique in the western world with Russia being the only other potential supplier of these isotopes. The HFIR is primarily dedicated to the neutron scattering sciences mission of the DOE Office of Science, but is able to produce these heavy isotopes while also satisfying its neutron scattering mission.

In summary, the number of research reactors available to produce radioisotopes worldwide is diminishing and those that remain are aging. There seems to be activity abroad to build new research reactors to fill future needs. However, the US fleet of aging research reactors remains constant for the time being. The U.S. URRs are in need of resources to refurbish and re-license. Only one reactor, MURR, is positioned to supply the medical and commercial needs of the country on a routine basis.

Table 3. Short-lived radioisotopes produced at small research reactors (<5MW)

Isotope	Half-life	Application
^{24}Na	15.0 hr	Tracer, Marker
^{28}Al	2.25 m	Student Laboratory
^{41}Ar	1.82 hr	Tracer
^{46}Sc	83.8 d	Tracer
^{76}As	26.3 h	Cancer therapy
^{79}Kr	1.4 d	Gas Tracer
^{82}Br	1.47 d	Tracer
^{90}Y	2.7 d	Cancer therapy
^{124}Sb	60.2 d	Tracer
^{133}Xe	5.24 d	Tracer, NPT Monitors
^{135}Xe	9.1 hr	NPT Monitors
^{140}La	1.68 d	Tracer, Gamma
^{192}Ir	73.8 d	Source Tracer
^{198}Au	2.3 d	Cancer therapy, marker

Table 4. Radioisotopes which can be produced at DOE research reactors

Isotope	Application
^{252}Cf	Neutron source – multiple uses
^{63}Ni	Pure beta source
^{238}Pu	Power source
^{60}Co	High specific-activity gamma source
^{177}Lu	Clinical trials
^{166}Ho/^{166}Dy	Targeted therapy, Ablation
^{166m}Ho	Calibration source
^{125m}Te	Diagnostic imaging
^{228,229}Th	Targeted therapy, Alpha emitter
^{227}Ac	Targeted therapy, Alpha emitter
^{99}Mo	Diagnostic imaging-Backup production only
^{64}Cu	Diagnostic imaging, Therapy
^{131}Ba/^{131}Ce	Cancer Brachytherapy
^{103}Ru/^{103m}Rh	Targeted therapy
^{186}Re	Targeted therapy
^{147}Pm	Power source
^{210}Po	Power source
^{170}Tm	Power source
^{171}Tm	Power source
^{242}Cm	Power source
^{35}S	Power source
^{144}Ce	Power source
^{75}Se	Well logging
^{225}Ac	Targeted Therapy, Alpha Emitter

The smaller URRs are currently able to meet the regional needs for short half-life isotopes for the time being, but the country cannot afford to lose many of these valuable resources. The DOE reactors are currently able to produce the unique high specific- activity isotopes and heavy actinides demands of the U.S., however, should the operation of the DOE reactors be interrupted, the only alternative source is Russia.

Accelerator Production of Radionuclides

Background

Accelerator isotopes are neutron deficient and are produced in either cyclotrons or linear accelerators by proton, deuteron, or alpha particle bombardment. Accelerator isotope applications generally complement reactor isotope applications, and accelerator isotopes usually decay by ₚ, , or positron emission or electron capture. Accelerator beam parameters, especially beam energy and beam current, are important considerations in the production of isotopes. Beam energy determines what isotopes are produced (and by what nuclear reaction) and beam current determines how much is produced. Low energy cyclotrons (<30 MeV) are generally used to produce short-lived isotopes (^{11}C, ^{15}N, and ^{18}F) that are used in clinical positron emission tomography (PET) and PET R&D. However, many other isotopes can be made at lower energies). Several commercial isotopes are produced in 30 MeV cyclotrons operated by industrial isotope producers and radiopharmaceutical manufacturers, e.g. ^{111}In, ^{201}Tl, ^{67}Ga, 123I and ^{103}Pd. Higher energy accelerators are usually operated by government laboratories and make products that require higher energy, e.g. ^{82}Sr.

Commercial Sources

Several commercial companies currently operate low and medium energy accelerators in the U.S. Many of these companies produce radioisotopes only for their own use. Other commercial companies sell the radioisotopes they produce to others. There are also many PET cyclotrons that are operated by joint ventures or partnerships with universities or hospitals.

Table 5. Commercial Radiopharmacies (CRP) and PET cyclotrons in the U.S.

	CRP	PET Cyclotrons
Cardinal Health NPS	182	20
Covidien	36	1
GE Healthcare	31	2
IBA Molecular	11	11
PETNET	42	42
Triad	28	7
Independent	96	23
Institutional[1]	69	6
Totals	495	112[2]

[1] The number of institutional PET cyclotrons that operate as CRPs.

[2] The total number of cyclotrons includes the 60 PET cyclotrons that are licensed as pharmacies are included in the total.

In many cases the hospital or university produces their own PET radiopharmaceuticals while the commercial partner markets any excess PET radiopharmaceutical capacity to other facilities. A number of these commercial capabilities are discussed below. Table 5 lists the commercial radiopharmacies (CRP) and PET cyclotrons in the US. (Note corporate ownership of these commercial entities has changed rapidly recently. The corporate structure and name identified may not be completely up to date.)

Lantheus Medical Imaging

In their North Billerica, MA facility Lantheus operates several medium energy cyclotrons. They produce ^{201}Tl and ^{67}Ga for use in their FDA approved radiopharmaceuticals. Lantheus is largely self sufficient for these radioisotopes and only relies on outside source of supply if there are operational issues with the cyclotrons such as an extended outage or un scheduled maintenance. Many of the other commercial radiopharmaceutical companies have made these radioisotopes available to each other if one of them has significant operational problems with their cyclotrons. This informal back up arrangement has worked well for many years. Lantheus does not routinely sell these radioisotopes in the open market.

Covidien

Covidien operates several medium energy cyclotrons at their facility in Maryland Heights, MO. They routinely produce ^{201}Tl, ^{123}I, ^{111}In, and ^{67}Ga for use in their FDA approved radiopharmaceuticals. On occasion they have also provided backup supply of these radioisotopes to the other radiopharmaceutical manufacturers. Covidien also operates more than 35 nuclear pharmacies. At a number of these nuclear pharmacies Covidien has partnered with a university or hospital to jointly produce 18_F FDG using low energy PET cyclotrons. This FDG is marketed to facilities located within a strategic distance to the nuclear pharmacy taking into account the short half life of ^{18}F.

MDS Nordion

MDS Nordion's Vancouver Operation has 3 cyclotrons (two high current TR30 cyclotrons, 30 MeV, and a single CP42 cyclotron, 42 MeV) that produce the standard commercially available radionuclides on a routine basis (^{201}Tl, ^{123}I, ^{67}Ga, ^{111}In, ^{103}Pd) as well as smaller amounts of ^{64}Cu for research centers. As with the other commercial vendors Nordion supplies other vendors during outages across the cyclotron sites.

GE Healthcare

GE Healthcare currently operates several medium energy accelerators in their Arlington Heights, IL and S. Plainfield, NJ facilities. They manufacture ^{201}Tl, ^{123}I, and ^{111}In in these cyclotrons for use in their own FDA approved radiopharmaceuticals. As with Covidien and Lantheus they do not routinely sell these radioisotopes commercially but have in the past served to provide backup supply top the other two major radiopharmaceutical manufacturers. GE healthcare also operates more than 30 nuclear pharmacies.

PETNET

PETNET, a Division of Siemens currently operates 42 PET cyclotrons in the U.S. They market ^{18}F FDG, ^{13}N NH$_4$, and provide some ^{11}C labeled compounds.

Cardinal Health

Cardinal Health currently operates nearly 182 nuclear pharmacies in the U.S. Twenty of these facilities are co-located with a PET cyclotron and are producing ^{18}F FDG.

IBA Molecular

IBA Molecular currently operates eleven PET cyclotrons throughout the Eastern U.S. These cyclotrons are used primarily for the production of ^{18}F FDG.

Independent Nuclear Pharmacies

There are more than 96 independent nuclear pharmacies located in the U.S. Some of these operate low energy cyclotrons for the production of ^{18}F FDG.

Trace Life Sciences/NuView

NuView in Denton, TX has two medium energy cyclotrons. They also have a 70 MeV LINAC which has been used at lower energy (33 MeV) to produce several medical radioisotopes. Trace has routinely operated four of the six target stations on the LINAC. The CP-42 has been operating for several years but the CS-30, which was relocated from Mount Sinai Medical Center has not been fully commissioned yet. The LINAC has been producing ^{201}Tl, ^{67}Cu, ^{64}Cu, ^{111}In, and ^{123}I for commercial distribution. Trace has submitted Drug Master Files (DMF) to the FDA for all of these products. They also have an aNDA for the production of Thallous Chloride, which has been distributed through NuView. Trace commercially distributes these radioisotopes for commercial use and for research and clinical studies.

DOE Capabilities

The discussion below focuses on the Department of Energy production capabilities on high energy accelerators at Brookhaven National Laboratory, Los Alamos National Laboratory, and international accelerator collaborators. Examples of isotopes produced at these facilities are given in Table 6.

Brookhaven National Laboratory Specific Capabilities

This program is part of the Medical Department at Brookhaven National Laboratory. It uses the *Brookhaven Linac Isotope Producer (BLIP), and the associated Medical Department laboratory and hot cell complex* to develop, prepare, and distribute to the nuclear medicine community and industry some radioisotopes that are difficult to produce or not available elsewhere. BLIP, built in 1972, was the world's first facility to utilize high energy protons for radioisotope production by diverting the excess beam of the 200 MeV proton LINAC to special targets. After several upgrades BLIP remains a world class facility and continues to serve as an international resource for the production of many isotopes crucial to nuclear medicine and generally unavailable elsewhere. The overall effort entails: (1) target design, fabrication and testing; (2) irradiations; (3) radiochemical processing by remote methods in the 9 hot cells of the Target Processing Lab; (4) quality control and analysis; (5) waste disposal; (6) facility maintenance; (7) new isotope and application development; and (8)customer liaison, marketing, packaging, and shipping. Service irradiations (without chemistry) are also performed.

Los Alamos National Laboratory Specific Capabilities

The **Los Alamos Neutron Science Center** is the cornerstone of Los Alamos isotope production. Historically, targets were irradiated at the beam stop at LANSCE from the inception of the facility in the 1970s. Irradiation of these targets with 800 MeV H+ protons produced isotopes by a nuclear process known as spallation. It became evident in the mid-1990's that continued delivery of H+ proton beam to the beam stop area would cease because of a lack of programmatic requirements. The Isotope Program proposed the construction of a new target irradiation facility that would divert beam from the existing H+ beam line in the transition region from the drift tube linac (DTL) to the side-coupled cavity linac (SCCL) into a new beam line and target station housed in a new facility adjacent to the existing accelerator facility. The energy of the protons in this transition region is 100 MeV, and production of isotopes in targets irradiated in this facility occur primarily by (p,xn) nuclear reactions. Approval was received for this proposal and the construction project was initiated in FY 1999. This new construction project, funded by DOE-NE, was completed in FY 2003 at a cost of $23.5 M. The 100 MeV Isotope Production Facility (IPF) has operated since the spring of 2004, and irradiates targets while LANSCE is operating for other experimental science programs. The 100 MeV IPF has also operated in a dedicated mode when target irradiations from other facilities are not available.

Table 6. Examples of radioisotopes produced with DOE accelerators and isotope applications

- Positron Emission Tomography (PET)
  - $^{82}Sr/^{82}Rb$ – myocardial imaging
  - $^{68}Ge/^{68}Ga$ – calibration sources for PET scanners, radiopharmaceutical research
  - $^{72}Se/^{72}As$ – oncological radiopharmaceuticals
- Isotopes for cancer therapy
  - ^{67}Cu – treatment of non-Hodgkin's Lymphoma
  - ^{103}Pd – seed implants for prostate cancer treatment
  - ^{76}As – bone pain palliation, radiopharmaceutical research for cancer treatment
- Environmental and research radiotracers
  - ^{32}Si – biological oceanography, global climate
  - ^{26}Al – acid rain, Alzheimer's research, materials
  - ^{95m}Tc – technetium behavior in ecosystems

The irradiated targets are transported from LANSCE in a shielded transportation container. The *TA-48 Hot Cell facility* at the Main Radiochemistry Site, Building RC-1 is the primary hot cell facility for accelerator isotope production. It consists of two banks of 6 chemical processing cells connected at one end by a large multipurpose "dispensary" cell, where all materials are received into and from which all materials leave the facility. Supporting facilities including several radiochemistry laboratories, a machine shop, two analytical laboratories, an extensive counting room facility, and offices for personnel surrounds the hot cell facility. This facility, along with the Laboratory's waste handling facilities, is absolutely essential for conducting the LANSCE isotope production mission.

University Based Production of Radionuclides

The university/hospital based accelerators are typically cyclotrons with proton energies below 20 MeV and are focused on internal programs for the production of the positron emitting isotopes such as ^{11}C, ^{15}O, and ^{18}F.

There are two principle suppliers of radionuclides for research in the university system: Washington University (St. Louis) School of Medicine and the University of Wisconsin. The University of Wisconsin program is fairly modest in scope but does provide 64Cu to several research groups in the U.S. The University of Buffalo operates a Cyclone-30 which has produced ^{18}F, ^{13}N, ^{11}C, ^{111}In, ^{62}Cu, ^{124}I and ^{15}O. Although they have a separate beam line and vault for target irradiation, they do not have hot cells capable of producing large quantities of radioisotopes. They are capable of producing quantities up to 1 Ci of ^{111}In per week, and had plans to submit a Drug Master File by the close of 2008. They have also produced batches of 110 mCi of ^{124}I per target.

The Washington University program has been supported for five years by the National Cancer Institute as a Research Resource to support Cancer researchers within the NCI system as well as other researchers needing radionuclides not readily available through commercial routes. This program has been extremely successful in providing ^{64}Cu as well as other radionuclides such as ^{76}Br, ^{86Y}, ^{94m}Tc, ^{66}Ga. The supply of radionuclides is based on two small cyclotrons, a CS-15 and a JSW, both accelerating 15 MeV protons. One of the mandates for this program was to transfer the technologies gained through this experience to a commercial partner. Washington University is in the process of doing this as of early 2009.

The Washington University experience illustrates that a small program can be effective if supported and that the program has a focus. This experience can serve as a model for other programs in support of radionuclide supplies.

There are more than 110 low and medium energy cyclotrons being operated in the U.S. largely for the production of ^{18}F FDG. Most of these cyclotrons are dramatically underutilized with less than 25% duty cycle. If one adds in the more heavily used commercial cyclotrons there are a total of close to 125 cyclotrons that have spare capacity and could be utilized for more production of research and commercialized radioisotopes.

3. RESEARCH OPPORTUNITIES IN BIOLOGY, MEDICINE, AND PHARMACEUTICALS

The majority of the isotopic material used in medical and biological research is used to support clinical trials. That said, there is still a significant demand for radioisotopes for use in research during radiochemical, in vitro, and in vivo preclinical investigations. These investigations are critical in the development of radiopharmaceuticals for diagnostic and therapeutic applications.

At the very earliest stage of development, a radionuclide is used to test labeling techniques that are used to radiolabel a targeting molecule. Obviously, given the cost of the starting material, the need to maximize the yield of the reaction is paramount. This is also important in maximizing the specific activity of the compound in question. In many cases, there is a limited number of sites on the target cell. In order to maximize the signal to noise ratio, or to maximize the therapeutic effect, it is important to have the greatest number of radioactive atoms attached to the target. These investigations are often limited to synthesis and analysis using standard techniques such as chromatography.

Once these investigations have been completed, testing of the biologic activity would be undertaken by utilizing cells that express the target that are exposed to the radiopharmaceutical. Usually, specific binding is determined by utilizing a control cell line of similar characteristics save the target itself and adding a substantial excess of the unlabeled targeting molecule to block specific binding to the target. In the case of a therapeutic conjugate, cell survival assays would also be undertaken to assess the cytotoxicity of the conjugate.

Once a radiopharmaceutical has passed these tests, testing in vivo will be done to assess the compounds ability to target the site of interest. In the case of human tumors, an animal with a compromised immune system will be used so that a xenograft of a tumor of interest can be grown. Usually, administration of the radiopharmaceutical will be via the vein as is the case in the nuclear medicine clinic. In the majority of cases, a combination of imaging of the radiopharmaceutical will be combined with post mortem tissue counting to determine relative and absolute uptake of the material.

Radioactivity can be selectively administered into patients by direct injection and if the radiation is in the right form, can selectively target human tumors and if sufficiently concentrated can destroy the tumors, without excessive damage to normal tissues. The prototype for this approach was introduced in the early 1940s, as radioactive iodine, in the form of ^{131}I which could target human thyroid cancer and in some cases cure patients of metastatic tumors which would have otherwise been fatal. Shortly after this, ^{32}P was introduced for targeting of human bone marrow disorders and was an early and quite effective therapy for abnormal states of myelodysplasia (pre-leukemia disorders) including polycythemia vera (abnormal increase in blood cells (primarily red blood cells) due to excess production of the cells by the bone marrow). These radionuclides were generally introduced in common chemical forms into the body, ^{32}P in phosphate form or ^{131}I as sodium iodide and because of natural processes within proliferating tissues achieve therapeutic concentrations. Although ^{32}P has been supplanted by more selective chemotherapies, ^{131}I continues to be the front-line drug for therapy of advanced thyroid cancer.

The practice of using radioactivity which is introduced into the patient by injection of relatively simple chemical forms continues to this day. For example Holmium and Dysprosium are injected into the joint space of patients for selective therapy of arthritis; ^{153}Sm, ^{89}Sr as simple chelates take advantage of the natural bone seeking properties of this class of chemical element. These two drugs have been approved by the Food and Drug Administration. Bone seeking elements that have slightly improved quality for palliative therapy are also being explored in clinical trials. These include ^{224}Ra and ^{110}Sn. Both are given as a simple salt, introduced in the patients with metastatic prostate cancer and are effective aviation therapies.

Targeted therapies with radio peptides or radio labeled antibodies have been introduced. Patients with non-Hodgkin's lymphoma are now routinely treated, especially in the late stage of their disease, with radio immunotherapy. This form of modern targeted therapies in medicine takes advantage of knowledge of the biology of cancer, and the specific biomolecules that are important in causing or maintaining the neoplastic state (abnormal proliferation of cells). In this case, an antibody or protein is used which is the carrier for the radioactivity, but so confers a specific binding property to a known component of any class of tumors: for example the radiolabeled peptide ^{90}Y (Yttrium DOTATOC) selectively binds to an endocrine receptor on carcinoid tumors, somatostatin type II (growth hormone inhibiting hormone). The targeting occurs much like a key (the radiopeptide) fitting into a lock (the somatostatin receptor), and over time sufficient radioactivity is deposited in the region of the tumor to damage the proliferating capacity of the tumor, in some cases eradicating sites of the tumor completely. In many instances clinical benefit is obtained from the use of the radioactivity, and especially in patients with advanced disease.

The use of therapeutic radionuclides is expanding in clinical research, and over the course of five years, it is likely that several additional FDA approved clinical applications will become best practice for specific clinical indications.

In general, it is in the particulate form of the radiation which is most likely to be useful for the purpose of depositing localized radiation in sufficient quantities to kill tumors without damaging normal tissues. Therapeutic radioisotopes are chosen for their radiation properties, including type of particulate radiation emitted, half-life, and energy. Radionuclides that are proposed for this type of therapeutic radiation usually emit one of three types of radiation: Auger electrons, beta particles, or alpha particles.

Auger electrons are emitted during the process of electron capture decay, and have the property of depositing energy very densely during the track of their decay. This process is referred to as high linear energy transfer (LET) of radiation to tissues. Obviously this process would be highly advantageous if a radionuclide were targeted to a tumor because then that energy would be deposited within and maximally damaging the tumor, sparing surrounding normal tissues. Beta particles are emitted from relatively neutron rich radionuclides that in general are produced in reactors as part of fission processes. The nuclear decay leads to the simultaneous emission of a neutrino, and a beta particle with a sharing of the available energy so that a description of the population of beta particles emitted has a distribution of decays for a given radionuclide characterized by more than simply the median and maximum energy. This is important for radiotherapy because the distance that a particle travels through tissue is proportional to its emission energy. Medium energy beta particles such as those from ^{131}I have a path length which is up to about 300 μ in tissues, while higher energy betas from radionuclides such as ^{90}Y have path lengths that may range up to 1 cm in tissue. This distribution of energies is considered undesirable because a significant portion of the deposited energy, especially for small tumors, will be deposited outside the tumor and in normal tissues. Alpha particles on the other hand are emitted with discrete energies, and because of their higher energy, slower velocities and higher charges deposit a large amount of energy along a relatively short track in tissues. It is generally considered that the cell nucleus is the killing zone within a cancer cell, and alpha particles traversing through a cell nucleus will deposit enough energy to kill the cell.

Different forms of radiation are useful for targeted therapy depending on the specifics of localization within a tumor or mass. For very small microscopic tumors, beta particles are less desirable because much of their energy would be deposited outside the tumor mass. On the other hand, if radionuclides bearing alpha particles could be targeted to cancer cells, it would take only a few radioactive decays to kill the cell.

As targeted vehicles including antibodies and peptides become more and more selective for selective binding to biomolecules attached to cancer cells, radionuclides which emit alpha particles have become more and more desirable. These radionuclides have been relatively difficult to get in sufficient quantities. The short-lived alpha emitters are particularly in demand, especially ^{225}Ac, ^{213}Bi, and ^{211}At.

Another area of compelling research with isotopes is in the development of pairs of isotopes in radiopharmaceuticals that can be used simultaneously for therapy and dosimetry. The therapeutic part of this research tests the ability of the radiation to either effectively ablate the tumor as determined by physical measurements or to "cure" the tumor. In order to better gauge the window of effectiveness and toxicity for the therapeutic agent, a surrogate agent is used. The second part of these new developments is the determination of the dosimetry of the compound. This information is then used to determine the dose that would be received by the target tumor and normal tissue without using the therapeutic agent itself. Obviously, the best option would be to use an isotope of the same element so that the chemical issues are the same. In Table 7, we present several examples of such therapeutic/dosimetry pairs.

Table 7. Pairs of isotopes that can be simultaneously used for dosimetry and therapy

Therapy mechanism	Therapeutic radionuclide	Diagnostic radionuclide for dosimetry	Decay mode of dosimetry agent
Beta decay	^{67}Cu	^{64}Cu	Positron
Beta decay	^{90}Y	^{86}Y	Positron
Beta decay	^{131}I	^{124}I	Positron
Alpha decay (daughter)	^{212}Pb (^{212}Bi)	^{203}Pb	Single Photon

For those compounds that pass these hurdles, patient studies will be undertaken either under the watch of a radioactive drug research committee and the institutional review board or after the investigators have applied to the Food and Drug Administration (FDA) for an Investigational New Drug (IND) status for the compound.

It should be evident that the amount of radionuclide required at each stage of development increases substantially. Thus, in parallel to the biomedical investigations underway, there needs to be a parallel effort to increase the amount of the radionuclide produced to support the research effort.

Identification of the Stable and Radioactive Isotopes that Are Needed to Realize These Opportunities

Stable Isotopes

Virtually all research studies of human *in vivo* metabolism today, in adults as well as children, employ stable rather than radioactive tracers. The movement away from radiotracers for such studies came over the last 35 years in large part due to the continued availability of stable isotopes from production programs at Los Alamos and Oak Ridge National Laboratories. The widespread use of ^{2}H, ^{13}C, and ^{18}O throughout basic and clinical biochemical research has made commercial production of these isotopes feasible and industry sources are readily available. ^{15}N demand is also met currently by industry sources, but it is not available domestically and there is no domestic generator of a new inventory. The latter is, potentially, no trivial problem because nitrogen is an indispensible dietary nutrient, especially in its role as the essential nutrient in amino acids, the building blocks of proteins. Thus, since there is no long-lived radiotracer alternative, an absence of 15$_N$ would curtail essentially all human studies of nitrogen metabolism. F, Na, Mg, P, S, Cl, K, Ca, Cr, Mn, Fe, Co, Ni, Cu, Zn, Se, Mo and I are essential nutrients in man. Some of these elements (e.g., F, Na, P, Mn, I) are mono-isotopic and, thus, not amenable for use in tracer studies. The remainder, however, have stable nuclides that are critically necessary for investigation of the requirements and metabolism of these indispensible nutrients in humans and animals.[2,3,4] Although these isotopes exist in current DOE inventory, the great bulk of the stable mineral isotopes used for human research are supplied by Russia and there is great concern for future availability. This concern has been expressed previously.[5]

It is not an exaggeration to say that research and clinical studies of essential mineral nutrient metabolism in man will come to a complete halt if the supply of these elements is curtailed. These concerns are no less acute or impactful in the domains of studying aquatic and terrestrial ecosystems where, in addition to the nuclides discussed above, the supply of stable isotopes of B, Cd, Ba, Hg, and Pb are, likewise, vitally essential for research into the impact of our environment on biological systems.[4]

Radioactive Isotopes

The radionuclide and the radiochemical purity of a given isotope of interest are critical. In the former instance, contaminating radionuclides can degrade the quality of the image, increase the dose to the patient, and render the product unusable according to the specifications for the radiopharmaceutical. In the latter instance, the chemical form of the material can potentially reduce the yield of the chemical reactions and potentially reduce the specific activity of the final product if a stable contaminant competes with the radionuclide during synthesis.

Estimated Quantity and Purity of Isotopes of High Priority for Biology, Medicine, and Pharmaceuticals

The opportunities identified in this area are listed in Table 8. The opportunities are listed in priority order for this section. Within each opportunity, if there is particular priority to one isotope, it is noted below. Most of these follow the recommendations of the "Report of Meeting Held to Discuss Existing and Future Radionuclide Requirements of the National Cancer Institute", held on April 30, 2008, the 2007 report of the National Research Council's Committee on the State of Nuclear Medicine, "Advancing Nuclear Medicine Through Innovation", and the list of projected isotope needs presented to the Committee by the National Cancer Institute from the on-going DOE-NIH working group.

Alpha therapies have extraordinary research potential and the isotopes of interest are ^{225}Ac, ^{211}At, and ^{212}Pb.The August isotope workshop identified the quantities of ^{225}Ac that would be needed for various stage clinical trails. One important factor for this isotope is that a potentially important interim source of ^{225}Ac is to recover the ^{229}Th parent from stores of ^{233}U that are scheduled to be diluted and disposed of, a process that would make them unsuitable for this purpose. It is estimated that a factor of four more material is needed at this point and if a Phase II study is undertaken, an order of magnitude more material will be needed. Because rapid action may be needed here, and the linking of ^{225}Ac with another isotope with the same parent, ^{213}Bi, it is given the highest priority in this opportunity.

The ^{211}At is needed in similar amounts, a factor of four more material now and an order of magnitude more should a Phase II study be undertaken. The ^{212}Pb availability is easier to expand than those of ^{225}Ac and ^{211}At since the grandparent ^{232}U has a shorter half-life and can be produced by neutron irradiation of ^{231}Pa.

Several low energy accelerators located at separate facilities in the United States are currently producing key medical research isotopes (^{64}Cu, ^{124}I). Other medical research isotopes (^{86}Y, ^{203}Pb, ^{76}Br, ^{77}Br) could also be produced at these accelerators. However, since these are research radionuclides and a large commercial market has not been established yet, operators of these accelerators do not have a significant incentive to produce these routinely. As a result, these radionuclides are not always readily available. There are significant advantages foreseen for sharing radiochemistry techniques and targetry technologies across accelerators located around the country in producing these research isotopes. These four diagnostic agents paired with theuraputic agents can all be made at all of these existing low energy accelerators, but to ensure regular and long term availability, there is a need for increased networking of producers and R&D in order to increase quantities needed by researchers. Within this opportunity, priority is not given to any individual pair of isotopes.

A continuously growing need for ^{89}Zr was projected by the DOE-NIH joint working group. This isotope is also produced at lower energy facilities than DOE currently operates and increased and regular availability requires coordination of production and the sharing of production and chemistry techniques. The production of ^{67}Cu requires higher energy accelerators than are currently avaiable at three sites, two of which are DOE NIPA facilities. The high demand projected for the future could not be met with current capacities. An isotope for medical applications should be considered a research isotope until it has been given New Drug Approval (NDA) by the FDA. Preclinical and clinical research subjects are administered these materials under guidance of a radioactive drug research committee or Investigational New Drug status. Until that time, the isotope should be considered as a research material and not subject to petitioning from a private provider. The experience to date has shown that premature abandonment of production has resulted in unsupported increase in price and a spotty ability to meet the demand of the research community.

Table 8. Research opportunities in Medicine, Pharmaceuticals and Biology in order of relative priority

Research Activity	Isotope	Issue/Action
Alpha therapy	^{225}Ac ^{211}At ^{212}Pb	Current sources are limited. One valuable source for ^{225}Ac, extraction of ^{229}Th from ^{233}U may soon be lost.
Diagnostic dosimetry for proven therapeutic agents	^{64}Cu ^{86}Y ^{124}I ^{203}Pb	Used in conjunction with ^{67}Cu therapy ^{90}Y therapy ^{131}I therapy and immune-diagnosis ^{212}Pb therapy The issue is the need for a coordinated network of production facilities to provide broad availability. There is need for R&D for common target and chemical extraction procedures.
Diagnostic Tracer	^{89}Zr	Immune-diagnosis 3.27 d half-life allows longer temporal window for imaging of MoAbs, metabolism, bioincorporation, stemcell trafficking, etc.
Therapeutic	^{67}Cu	Requires specialized high energy production facilities and enriched targets

4. RESEARCH OPPORTUNITIES WITH ISOTOPES IN PHYSICAL SCIENCES AND ENGINEERING

Isotopes give unique responses under various excitations in solids, liquids, and gases. This is either due to the mass difference, which couples to electronic degrees of freedom, or nuclear structure, which shows large variations even with a single neutron addition. This unique behavior of isotopes lends itself to a plethora of useful applications. Thus, in almost all branches of sciences and engineering, from humanities to environmental science to nuclear physics and geology, isotopes have found fundamental and technological applications. For example, isotopes have changed the way we produce energy, develop industrial diagnostic methods, learn about our past in sociological (archeology), geological (terrestrial and extra-terrestrial), ecological (carbon and nitrogen cycle), and astronomical sense, help us secure our future energy needs, manage our natural resources like water and forests, and provide home and food safety. Therefore, many aspects of isotope production, use, detection, and education for research are relevant and worthy of continued support by DOE.

While the discovery of isotopes is less than 100 years old, today we are aware of about 250 stable isotopes of the 90 naturally occurring elements. The number of natural and artificial radioactive isotopes exceeds 3200, already, and this number keeps growing every year. F. Soddy's discovery (1910) of lead (Pb) obtained by decay of uranium and thorium differing in mass was considered a peculiarity of radioactive materials. In 1913 Soddy, and independently Fajan, developed a displacement law, which explains the change in mass and in the place in the periodic table after -decay or -decay takes place, and extended its implications on the formation of isotopes.

In this chapter, we highlight several research opportunities with isotopes across various fields. Isotopes are essential tools in basic research across all of *nuclear physics*. Many of the most important experiments at the frontiers depend on reliable and affordable availability of nuclear isotopes. In understanding the nucleon at the fundamental quark and gluon level, targets and beams of ^{2}H and ^{3}He allow access to the neutron. In looking beyond the Standard Model with tests of fundamental symmetries, the important experiments rely on a number of key isotopes. Finally, one of the central thrusts of modern nuclear physics is to understand the structure and properties of rare isotopes – those that exist only for a short time but which play a central role in the formation of the elements. A new user facility (FRIB) is planned in the U.S. to address this important area.

Basic Principles, Intrinsic Characteristics, and Fundamental Applications of Isotopes

In *nuclear physics*, there are many uses of enriched isotopes, stable and radioactive. Enriched stable isotopes are needed for targets and for accelerated beams at various laboratories producing both stable and radioactive beams needed to study the structure of nuclei. For example, ^{48}Ca is a very neutron rich isotope that is commonly used as a beam at various nuclear physics laboratories to study the properties of exotic nuclei far from stability. Also, it is used in fragmentation reactions to produce very exotic radioactive beams. A future supply of stable highly enriched isotopes of many different elements is necessary for forefront experiments in nuclear physics. Below we list some examples of frontier research experiments with special needs in isotopes.

The Argonne Tandem Linac Accelerator System (ATLAS) is a DOE-funded national user facility for the investigation of the structure and reactions of atomic nuclei in the vicinity of the Coulomb barrier. A major advance in rare-isotope capabilities at ATLAS will be the *Californium Rare Ion Breeder Upgrade (CARIBU)*. Rare isotopes will be obtained from a one-Curie ^{252}Cf (Californium) fission source located in a large gas catcher from which they will be extracted, mass separated, and transported to an Electron Cyclotron Resonance (ECR) source for charge breeding prior to acceleration in ATLAS. This will provide accelerated neutron-rich beams with intensities up to 7×10^5/s, and will offer unique capabilities for a few hundred isotopes, many of which cannot be extracted readily from Isotope Separator On Line (ISOL) type sources. In addition, it will make these accelerated beams available at energies up to 10-12 MeV/nucleon, which are difficult to reach at other facilities. At the present time, the availability of ^{252}Cf for this purpose is in question, in part due to competing commercial and security demands for this DOE-produced isotope.

An alternate and very powerful probe of new electroweak CP violation is to search for a *permanent electric dipole moment (EDM)* of an elementary particle or quantum bound state. The principles of quantum mechanics tell us that the interaction between an EDM and an applied electric field E is proportional to $S \cdot E$, where S is the spin of the particle or quantum system. This interaction is odd under both time reversal (T) and parity (P) transformations. By the CPT theorem of quantum field theory, a nonzero EDM implies the presence of CP violation. Other very promising experiments are also under development to search for atomic and electron EDMs. Certain radioactive atoms possessing a large octupole deformation are expected to have greatly enhanced sensitivity to CP-violating forces in the nucleus. Both ^{225}Ra and ^{223}Rn show promise as potential high-sensitivity deformed nuclei. Currently experiments using these nuclei are being planned or pursued at laboratories around the world, including Argonne National Laboratory (using ^{225}Ra extracted from a ^{229}Th source at ORNL) and TRIUMF in Canada (using a radioactive beam). The precision of the ^{225}Ra experiment is projected to be limited by the isotope supply.

Neutrinoless double beta (0νββ) decay experiments could determine whether the neutrino is its own antiparticle, and therefore whether nature violates the conservation of total lepton number: a symmetry of the Standard Model whose violation might hold the key to the predominance of matter over antimatter. Multiple *0νββ* experiments using different isotopes and experimental techniques are important not only to provide the required independent confirmation of any reported discovery but also because different isotopes have different sensitivities to potential underlying lepton-number-violating interactions.

CUORE - *the Cryogenic Underground Observatory for Rare Events* - is a bolometric detector searching for *0νββ* in ^{130}Te. The Italian–Spanish–U.S. collaboration plans to install and operate TeO$_2$ crystals containing 200 kg of ^{130}Te at the underground Laboratori Nazionali del Gran Sasso in Italy. Replacing the natural Te with isotopically enriched material in the same apparatus would subsequently lead to a detector approaching the ton scale.

The Majorana collaboration is engaged in a research and development effort to demonstrate the feasibility of using hyperpure germanium (Ge) diode detectors in a potential one-ton-scale *0νββ* experiment. The initial Majorana research and development effort, known as the Majorana Demonstrator, utilizes 60 kg of Ge detectors, with at least 30 kg of 86% enriched ^{76}Ge in ultralow background copper cryostats, a previously demonstrated technology. This Canadian–Japanese–Russian–U.S. collaboration is in close cooperation with the European GERDA Collaboration, which proposes a novel technique of operating Ge diodes immersed in liquid argon. Once the low backgrounds and the feasibility of scaling up the detectors have been demonstrated, the collaborations would unite to pursue an optimized one-ton-scale experiment.

Several other promising opportunities to carry out sensitive *0νββ* experiments exist, and U.S. nuclear physicists have indicated an interest in being involved. One notable experiment is known as SNO+, a proposed ^{150}Nd-doped scintillator measurement that would utilize the previous Canadian hardware of the acrylic sphere, photomultiplier tubes, and support system—engaged in a coordinated international program of *0νββ* measurements.

An isotope that is broadly used in nuclear physics as well as low temperature physics is ^{3}He. In addition, ^{3}He is widely used as a neutron detector. In particular, polarized ^{3}He is widely used as an effective polarized neutron in scattering experiments, e.g., at Jefferson Lab. There are plans to implement a polarized ^{3}He source at BNL to provide polarized neutron beams at the Relativistic Heavy Ion Collider (RHIC). A beam of polarized ^{3}He is also a central element in the neutron EDM experiment, planned for the SNS.

Many unusual phases of matter like superfluidity, superconductivity, and Bose-Einstein condensation occur at extremely low temperatures, which enable study of subtle behaviors that are obscured by thermal motion at higher temperature. To reach a temperature below 0.3

K, one would need the ^{3}He-^{4}He dilution refrigerator, because it can operate continuously, provide a substantial cooling power at temperatures from around 1.0 K down to 0.010 K and below, and it can run uninterrupted for months. The ^{3}He-^{4}He dilution refrigerator is invaluable for experiments that require temperatures as low as 0.001 K because it can be used to pre-cool the adiabatic demagnetization systems.

There are many other scientific areas that require enriched isotopes. Mass differences between different isotopes cause sufficient change in bond strength and vibrational characteristics of volatile compounds of H, C, N, and O to affect their heat of vaporization. Thus, time, temperature, and geographical variations of isotope ratio differences can be used as a tracer of climate change, and help quantify the hydrogen, carbon, nitrogen, and oxygen cycle on earth. Sources of isotopes are essential as calibration standards.

For example, in *Paleoclimatology*, which studies climate change over the entire history of the Earth, oxygen isotope ratios[6] play an important role. Water with oxygen-16, $H_2^{16}O$, evaporates at a slightly faster rate than $H_2^{18}O$; this disparity increases at lower temperatures. The $^{18}O/^{16}O$ ratio provides a record of ancient water temperature. The measured heat capacity difference between H_2 ^{18}O and H_2 ^{16}O is 0.83 ± 0.12 J K^{-1} mol^{-1} for liquid water.[7] When global temperatures are lower, snow and rain from the evaporated water tends to be higher in ^{16}O, and the seawater left behind tends to be higher in ^{18}O. Marine organisms would then incorporate more ^{18}O into their skeletons and shells than they would in warmer climates. Paleoclimatologists directly measure this ratio in the water molecules of ice cores, or the limestone deposited from the calcite shells of microorganisms. Calcite, $CaCO_3$, takes two of its oxygen from CO_2, and the other from the seawater. The isotope ratio in the calcite found in the skeletons and shells of marine organisms is therefore the same as the ratio in the water from which the microorganisms of a given layer extracted, after readjusting for CO_2.

Nitrogen isotopic ratios also provide a powerful tool for evaluating processes within the nitrogen cycle and for reconstructing changes in the cycling of nitrogen through time. The biologically-mediated reduction reactions that convert nitrogen from nitrate (NO_3^{-1},+5 oxidation state) to nitrite (NO_2^{-1}, +3) to nitrous oxide (N_2O^{+1}), to nitrogen gas (N_2^{0}), and to ammonia (NH_3^{-3}) are faster for ^{14}N than for ^{15}N as a result of higher vibrational frequency of bonding to ^{14}N than to ^{15}N .This results in products that are ^{15}N-depleted relative to the substrate. If the substrate reservoir is either closed or has inputs and outputs that are slow relative to one of the reduction processes then the reservoir will become enriched in ^{15}N. Therefore, the stable isotope ratio of nitrogen can be a promising proxy for delineating the eutrophication in the environment, which is a process describing an increase in chemical nutrients — compounds containing nitrogen or phosphorus — in an ecosystem. Since nitrogen is one of the important nutrient elements in a lake and abundant in anthropogenic sewage and chemical fertilizers, a range in fractionations of nitrogen isotope ratios in aquatic processes makes nitrogen isotope ratios an excellent tracer to monitor eutrophication.

In *astrophysics and planetary sciences*, measurements of D/H, $^{13}C/^{12}C$, $^{15}N/^{14}N$, or $^{18}O/^{16}O$ of primitive solar system materials record evidence of chemical and physical processes involved in the formation of planetary bodies and provide a link to materials and processes in the molecular cloud that predated our solar system. Modern developments exploiting nano-SIMS method have provided mineralogical and isotopic evidence of origins of stardust as composed of precursors of the solar system (McKeegan, et al, Science 314

(2006) 1724). Again, the isotope production requirements here are for measurement standards.

In *solid-state physics*, vibrational spectroscopy methods such as Brillouin light scattering, or Raman spectroscopy, plays a major role in using "isotope labeling", in applications such as identifying the origins of meteorites, or magnitude of atomic displacements in a complex molecule. In superconductivity, shift in transition temperatures with isotopic substitution is a well-established approach to understand the mechanisms of formation of Cooper pairs, and their physical location inside complex crystals. Presence of mixed isotopes also acts as scattering centers in an otherwise perfect crystal, reducing cooperative behavior of atoms with substantially reduced thermal conductivity. Nuclei with unpaired spins can couple with electron spins, and the difference in decay time lends nuclear spin as a solid-state quantum memory. Isotopically enriched silicon or germanium-based semiconductors lend themselves for engineered nanostructures with phase coherence quality suitable for solid-state quantum memory devices. In chemistry, elusive transition states in reaction chemistry can be revealed through isotopic labeling. In exploiting the variations in a nuclear energy level between different isotopes lead to isotope-based spectroscopic methods, such as *Mössbauer spectroscopy*, which is a major research tool across many scientific disciplines. For example, decay of ^{57}Co, through an electron capture process to ^{57}Fe, provides an ideal parent/daughter relationship that lends itself to study in hyperfine interactions in magnetism, lattice dynamics, and local atomic structure in condensed matter in an unprecedented energy resolution of 10^{-13} or better. Over 50,000 papers have been published in Mössbauer spectroscopy, and a total of 114 isotopes have been used. Today many of the parent/daughter isotopes are available only from a single country (not the U.S.), which is a cause for concern for the scientific community. Mössbauer isotopes are typically produced either in a cyclotron via deuterium bombardment or in a reactor.

In determining *fundamental constants and metrology*, developing a mass standard in fundamental units has been a struggle. The current approach, dubbed Avogadro's project, is an ongoing international collaboration between laboratories in Germany, Italy, Belgium, Japan, Australia, and USA to redefine the kilogram in terms of the Avogadro constant. The Avogadro constant is obtained from the ratio of the molar mass to the mass of an atom, and it is known to an uncertainty of 0.1 ppm. The goal is to reduce this to 0.01 ppm by measuring the volume and mass of isotopically pure silicon spheres. For a crystalline structure such as silicon, the atomic volume is obtained from the lattice parameter and the number of atoms per unit cell. The atomic mass is then the product of the volume and density. The limiting factors are the variability from sample to sample of the isotopic abundances of Si and the content of impurities and vacancies. Thus kilograms of isotopically pure ^{28}Si are needed, which is only provided by Russia. Currently two such 1 kg spheres are available. The new spheres were made from just one isotope: ^{28}Si. The mono-isotopic silicon was made in Russia while the near perfect crystal was grown in Germany, and perfect spheres were cut in Australia. To achieve the required concentration of the ^{28}Si isotope, a new centrifugal method was used for producing stable isotopes. SiF_4 of natural isotopic composition was used as a compound for centrifugal enrichment of ^{28}Si. A special centrifugal setup and a technology for production of $^{28}SiF_4$ with extremely high concentrations were developed in the Tsentrotekh-EKhZ Science and Technology Center. As a result, $^{28}SiF_4$ with an isotopic purity of 99.992–99.996% was produced.[6]

Table 9. Research Opportunities in Physical Science and Engineering in order of relative priority

Research activity	Isotope	Issue/action
Begin new facility to produce and study radioactive beams of nuclei from ^{252}Cf fission, for research in nuclear physics and astrophysics - CARIBU at ANL	^{252}Cf (2.6 yr)	Supply of ^{252}Cf is uncertain; 1 Ci source is needed each 1 1/2 year for at least four years.
Measure permanent atomic electric dipole moment of ^{225}Ra to search for time reversal violation, proposed to be enhanced due to effects of nuclear octupole deformation;	^{225}Ra (15 d)	Supply of ^{225}Ra is limited. Need 10 mCi source of ^{225}Ra every two months for at least two years
Create and understand the heaviest elements possible, all very short-lived and fragile. Study the atomic physics and chemistry of heavy elements for basic research and advanced reactor concepts.	^{209}Po, ^{229}Th,^{232}Th, ^{231}Pa, ^{232}U, ^{237}Np,^{248}Cm, ^{247}Bk	Make certain actinides in HFIR and then prepare targets for accelerator-based experiments to make superheavy elements; targets needed are ^{241}Am, ^{249}Bk, ^{254}Es - not available now; need10 - 100 mg on a regular basis; purity is important
Neutron detectors, electric dipole moment measurement, low temperature physics,	^{3}He	Total demand exceeds that available
Isotope dilution mass spectrometers	^{236}Np, ^{236,244}Pu, ^{243}Am, ^{229}Th	High purity ^{236}Np is not available; others are in limited supply; 10 - 100 mg needed on a regular basis; purity is important
Search for double beta decay without neutrino emission - an experiment of great importance for fundamental symmetries	^{76}Ge	Need to fabricate large detectors of highly enriched ^{76}Ge; U.S. cannot produce quantity needed, ~1000 kg
Spikes for mass spectrometers	^{202,203,205}Pb, ^{206}Bi, ^{210}Po	202,205$_{Pb}$ difficult to get in high purity in gram quantities
Avogadro project - worldwide weight standard based on pure ^{28}Si crystal balls	^{28}Si	Concern about future supply and cost of kg of material needed
Radioisotope micro-power source	^{147}Pm, ^{244}Cm	Development needed for efficient conversion
Isotopes for Mossbauer Spectroscopy, over 100 radioactive parent/stable daughter isotopes	^{57}Co, ^{119m}Sn ^{67}Ni, ^{161}Dy, ...	Some Isotopes only available from Russia, a concern for scientific community

A very practical but important power-source type application is *radioisotope thermoelectric generators (RTG)*. Usage of RTG batteries can be very esoteric and unique. For example, they have been used as power source for spacecrafts (*Apollo, Pioneer, Viking, Voyager, Galileo, Cassini*), where a few hundred watts of power is needed for a very longtime. They can also be used in very practical and large-scale applications like driving pacemakers and other implanted medical devices, where microwatts of power are needed. Various technologies are under development including stirling heat engines (devices that convert heat energy into mechanical power by alternately compressing and expanding a fixed quantity of air or other gas (the *working fluid*) at different temperatures), thermo-photovoltaic devices using piezoelectric materials combined with MEMS (micro electro mechanical systems) technology. The most suitable isotope for RTG applications is ^{238}Pu. It is an alpha emitter, thus it has the lowest shielding requirements and long half-life (87.7 years) high density (19.6 g/cc) and reasonably high energy density (0.56 W/g). While there are concerns for environmental and other safety concerns, potential improvements in energy efficiency and prevention of radiation damage for some piezoelectric converters may increase the electrical conversion efficiency by a factor 10 or more, thus making RTGs very attractive power sources and, in some cases, maybe the only alternative. Therefore, the need for alpha emitting isotopes of ^{238}Pu, ^{244}Cm, ^{241}Am, and beta-decaying ^{90}Sr will continue in the future.[7]

Table 9 lists the identified research opportunities, ordered by priority in the physical sciences and engineering areas. Within each opportunity, no particularly ordering of priority for individual isotopes has been assigned when more than one isotope is mentioned. The prioritizations are based on our own expertise and the priorities presented to NSACI from the DOE-ONP and DOE-BES programs. The relative priorities of items in Table 9 are discussed in the recommendations section. The lighter tone of blue in Table 9 highlights the relatively higher research opportunity potential of these topics. For example, the first four items of this table are the substance behind recommendations 3 and 4 of this report; "...the creation of a plan and investment in production to meet these research needs for heavy elements" and "...a focused study and research & development to address new or increased production of ^{3}He", respectively. The dark blue items are addressed in recommendation number 5 in support of research and development activities towards re-establishing a domestic capability for mass separated stable and radioactive isotopes.

5. RESEARCH OPPORTUNITIES WITH ISOTOPES FOR NATIONAL SECURITY AND OTHER APPLICATIONS

Isotopes are used in many areas related to nuclear security. DHS, NNSA, and the FBI require radioisotopes for the calibration and testing of instrumentation used for the analysis of nuclear materials. NNSA also performs nuclear physics measurements that utilize radioisotopes for these calibration and testing purposes. In addition these organizations use enriched stable isotopes for calibration and isotope dilution measurements in mass spectroscopy. All of these activities require relatively small amounts of these materials and there have been no difficulties in supplying these needs. In addition to these operational needs for sources there are some other activities that require larger quantities of materials or materials that are more difficult to obtain.

DHS is currently deploying many large radiation detection systems to monitor cargo that enters the United States. These devices measure both gamma-rays and neutrons and use this information to detect the presence of nuclear materials. The neutron detectors in these devices use ^{3}He tubes; this type of detector has excellent stability and high efficiency for detecting neutron radiation from plutonium. DHS plans on deploying a large number of these detectors in the course of the next five years. In a similar manner the "Second Line of Defense" program in NNSA/NA-25 will also deploy a large number of these same types of detection systems in foreign ports that ship cargo to the U.S. It would appear that the demands of these two programs will exhaust the U.S. reserve of ^{3}He and require more than the domestic annual production of this stable isotope.

However, this is not the only use of ^{3}He. As mentioned previously, the field of low-temperature physics uses ^{3}He in their dilution refrigeration systems that make all of their work possible. Also, polarized ^{3}He is used for magnetic resonance imaging for lung scans. *Therefore, this apparent shortfall of ^{3}He will not only hinder national security programs but will also have a devastating effect on the research and medical activities that rely on this isotope.*

In addition to the deployment of large detector systems, DHS is also conducting research on the effects of nuclear devices and "dirty bombs". This research uses radioisotopes of sufficiently long half-lives that are similar to those of interest to study the effects of radiation exposure to humans and the environment. Radiotracers with shorter half-lives have more easily detectable emissions and are preferable for research studies of their biologic or environmental disposition due to less potential waste problems. The research studies in this area will determine the most effective method to decontaminate radioisotopes in the environment. This research will also help in the development of agents that could remove radioisotopes from an exposed individual.

DTRA, DHS, and NNSA are also involved in the nuclear forensics of a possible domestic nuclear event. The analyses involved with such forensics activities require the use of both stable and radioisotopes as tracers, such as ^{236}Pu, for the complex radiochemistry separations that are used in this analysis. For this application the quantities of the needed isotopes are modest and can be met by the existing inventories for the next several years.

The area of weapons physics also requires the use of isotopes. With the cessation of nuclear testing, the challenge for the national nuclear security program has been to certify the safety and reliability of the enduring stockpile. Central to this was the realization that the "parametric" engineering based development program that historically served the program well would have to be modified to have increased emphasis on a more fundamental scientific understanding of weapon performance. With the development of the modern ASCI based supercomputer capabilities it has become possible to computationally investigate the evolution of a nuclear explosion at an unprecedented level. However, this procedure will only result in a reliable predictive capability if commensurate effort is expended to insure that correct underlying physical data is used in the codes.

The nuclear processes occurring in the explosion are the fundamental heart of the device. A correct understanding of the nuclear reactions and their resultant radiation and particle transport must be accomplished. To this end, nearly every test has utilized "radchem" detectors to provide spatially resolved information on the device performance. The archived data from these tests represents a treasure of detailed information that can provide improved understanding of the underlying weapon physics. These radchem detectors have often been used to diagnose the 14 MeV neutrons produced in the thermonuclear reactions. In the high neutron fluence environment of a nuclear device, multiple nuclear reactions can occur on single radchem detector atoms. These higher order reactions often occur on radioactive isotopes for which little, if any, experimental data exist for their reaction cross sections. Since the radchem production is analyzed at times long compared with the explosion process, these materials are exposed to the complete integral fluence of the produced neutrons. In particular, as the neutrons evolve during the explosion dynamics they are down-scattered in energy eventually approaching some local environmental thermodynamic equilibrium. At these lower energies the dominant reaction becomes neutron capture. These "late time" effects can result in a perturbation of the isotopic abundances produced in the early thermonuclear burning of the device.

The interpretation of the device-produced isotopic yields is highly dependent on nuclear modeling. Though great improvements in the understanding of nuclear reactions have been made over the years, the a priori prediction of neutron capture cross sections remains very difficult. We could obtain improved data for capture cross sections on unstable species an experimental program has been launched that uses unique LANL capabilities. These include: (1) neutrons produced at Lujan Center at LANSCE; (2) a new detector system called DANCE (Detector for Advanced Neutron Capture Experiments) - a 4it 140 element BaF2 detector array to measure capture reactions: (3) capabilities for radiochemical processing of irradiation materials; and (4) a dedicated isotope separator (RSIS, Radioactive Sample Isotope Separator located in the CMR building) of radioactive species for target preparation. To complete this integral LANL program it is necessary to have a capability to produce the isotopes required for these measurements. The newly commissioned Isotope Production Facility can play a critical role in providing these required isotopes. This research program would therefore provide useful data for weapons physics as well as develop capabilities and experts in the area of isotope production and nuclear science.

While there are no high priority research opportunities identified in this security applications area, the following observations apply more broadly across the entire NIPA program.

- Nuclear security needs will exhaust our supply of ^{3}He, we recommend that DOE/NNSA and DHS should consider alternative materials or technologies for their neutron detectors.
- The Nuclear Security applications of isotopes will always benefit by those programs that maintain our domestic capabilities to produce isotopes.
- National Security interests would be served by the development of a domestic source for a wide range of stable isotopes rather than relying on sources in Russia.
- There is a growing need for more experts in radiochemistry and other technical areas related to isotope production. DOE needs to help universities produce more of these experts.
- DOE should be able to characterize the isotopes that they produce with respect to nuclear forensics.

6. RECOMMENDATIONS FOR CHARGE 1

Compelling research opportunities were identified and presented in prioritized lists within the two areas of 1) biology, medicine, and pharmaceuticals, and 2) physical sciences and engineering. The third area 3) security applications did not have immediate research priorities but made a number of observations and recommendations that apply more broadly for the entire NIPA program. While it is challenging to assess relative scientific merit across disciplines, we have identified the highest priorities for the most compelling research opportunities. These recommendations also define the relative priorities of opportunities in Tables 8 and 9.

There are compelling research opportunities using alpha-emitters in medicine. There is tremendous potential in developing far more effective treatments of cancers by the use of alpha-emitters in comparison to other radio-isotopes. Therefore, development and testing of therapies using alpha emitters are our highest priority for research isotope production for the medical field. This priority is reinforced by the potential need for rapid action due to the 2012 deadline for downblending of current DOE stocks of ^{233}U, a procedure that would eliminate its value as a source of ^{225}Ac.

1. Invest in new production approaches of alpha-emitters with highest priority for ^{225}Ac. Extraction of the thorium parent from ^{233}U is an interim solution that needs to be seriously considered for the short term until other production capacity can become available.

There is strong evidence for the potential efficacy of pairs of isotopes with simultaneous diagnostic/therapeutic capabilities. Table 8 of this report presents a prioritized list isotopes that have the greatest research potential in Biology, Medicine, and Pharmaceuticals. NSACI finds the research opportunities offered with these pairs of isotopes to be the second highest priority in identifying compelling research opportunities with isotopes. Many of these isotopes could be produced at existing accelerator facilities. We recommend the maximization of the production and availability of these isotopes domestically in the U.S. through investments in research and coordination between existing accelerators. The panel felt that such a network could benefit all areas of basic research and applications from security to industry. This should include R&D to standardize efficient production target technology and chemistry procedures.

2. We recommend investment in coordination of production capabilities and supporting research to facilitate networking among existing accelerators.

The basic physical sciences and engineering group prioritized research opportunities across various disciplines and a summary of this prioritization is given in Table 9. The availability of californium, radium, and other trans-uranic isotopes, the first three opportunities in Table 9, are particularly important for research.

3. We recommend the creation of a plan and investment in production to meet these research needs for heavy elements.

Experts in the nuclear security and applications areas strongly consider the vulnerability of supply from foreign sources to be of highest priority. This concern was echoed strongly by all members of the subcommittee in from medicine to basic science and engineering. Additionally, the projected demand for ^{3}He by national security agencies far outstrips the supply. This would likely endanger supply for many other areas of basic research. While it is beyond our charge, it would be prudent for DOE/NNSA and DHS to seriously consider alternative materials or technologies for their neutron detectors to prepare if substantial increases in ^{3}He production capacity cannot be realized.

4. We recommend a focused study and R&D to address new or increased production of ^{3}He.

The remaining isotopes in Tables 8 and 9 all are promising research opportunities, and funds for production from the Research Isotope Development and Production Subprogram would be well spent on targeted production of these isotopes to meet immediate research needs, especially if unique production opportunities arise. However, at this point in prioritization, NSACI concludes that larger, long-term issues should take priority. The darker tone of blue used in Table 9 is an indication of that.

An important issue for the use of isotopes is the availability of high-purity, mass-separated isotopes. The stable isotopes ^{76}Ge and ^{28}Si (^{3}He is stable but obtained from the beta-decay of ^{3}H, not by isotope separation) listed in Table 9 are needed in large quantities that present special problems. While no other individual stable isotope reached the level of the highest research priority, the broad needs for a wide range of mass-separated isotopes and the prospect of no domestic supply raised this issue in priority for the subcommittee. NSACI feels that the unavailability of a domestic supply poses a danger to the health of the national research program and to national security. NSACI recommends:

5. Research and Development efforts should be conducted to prepare for the reestablishment of a domestic source of mass-separated stable and radioactive research isotopes.

Vital to the success of all scientific endeavors is the availability of trained workforce. While the scientific opportunities have expanded far beyond the disciplines of radiochemistry and nuclear chemistry, the availability of trained personnel remains critical to the success of research in all frontiers of basic science, homeland security, medicine, and industry. The individual research areas must make concerted efforts to invest in work-force development to meet these needs. The isotope program has a special responsibility to ensure a trained workforce in the production, purification and distribution of isotopes.

6. We recommend that a robust investment be made into the education and training of personnel with expertise to develop new methods in the production, purification and distribution of stable and radio-active isotopes.

All of the issues and recommendations considered here will be important input for answering the 2nd NSACI charge (See Appendix 1) due in 31 July 2009, developing a long range plan for the Nuclear Isotopes Production and Application Program.

REFERENCES

Abrams SA, Klein PD, Young VR, Bier DM. Letter of concern regarding a possible shortage of separated isotopes. *J Nutr* 122:2053 (1992).

Doklady Chemistry, 2008, Vol. 421, Part 1, pp. 157–160. © Pleiades Publishing, Ltd. (2008).

Fairweather-Tait SJ and Dainty J. Use of stable isotopes to assess the bioavailability of trace elements: a review. *Food Addit Contamin* 19:939-947 (2002).

For example, the most popular Mössbauer radioactive parent isotope, ^{57}Co, is produced by irradiating an iron target with 9.5 MeV deuterons following ^{56}Fe(d,n)^{57}Co. After irradiation, the target is dissolved in mineral acids, followed by isopropyl ether extraction and an ion exchange separation. The next most widely used isotope, ^{119m}Sn, is produced through ^{118}Sn(n,y)^{119m}Sn reaction or by electron capture decay of ^{119}Sb obtained from ^{119}Sn(p,n) ^{119}Sb or ^{120}Sn(2p,n) ^{119}Sb reaction (Spectroscopy Handbook, Ed. J. W. Robinson, CRC Press (1981)).

J. Norenberg, P. Stapples, R. Atcher, R. Tribble, J. Faught and L. Riedinger, Report of the Workshop on The Nation's Need for Isotopes: Present and Future, Rockville, MD, August 5- 7, 2008, http://www.sc.doe.gov/henp/np/program/docs/Workshop%20Report_final.pdf

Nagano, et al, *J. Phys. Chem. 97,* 6897-6901 (1993).

Robert D. Koudelka, Radioisotope Micropower System Using Thermophotovoltaic Energy Conversion AIP Conf. Proc., Volume 813, pp. 545-551 (2006).

Stürup S, Rüsz Hansen H, Gammelgaard B. Application of enriched stable isotopes as tracers in biological systems: a critical review. *Anal Bioanal Chem* 390:541-554 (2008).

Turnland JR. Mineral bioavailability and metabolism determined by using stable isotope tracers. *J Anim Sci* 84 (Suppl): E73-E78 (2006).

APPENDIX 1: THE NSAC CHARGE

August 8, 2008

Professor Robert E. Tribble
Chair, DOE/NSF Nuclear Science Advisory Committee
Cyclotron Institute
Texas A&M University
College Station, TX 77843
Dear Professor Tribble:

The Fiscal Year (FY) 2009 President's Request Budget proposes to transfer the Isotope Production Program from the Department of Energy (DOE) Office of Nuclear Energy to the Office of Science's Office of Nuclear Physics, and rename it the Nuclear Isotope Production and Applications program. In preparation for this transfer, this letter requests that the Nuclear Science Advisory Committee (NSAC) establish a standing committee, the NSAC Isotope (NSACI) sub-committee, to advise the DOE Office of Nuclear Physics on specific questions concerning the National Isotope Production and Applications (NIPA) Program. NSACI will be constituted for a period of two years as a subcommittee of NSAC. It will report to the DOE through NSAC who will consider its recommendations for approval and transmittal to the DOE.

Stable and radioactive isotopes play an important role in basic research and applied programs, and are vital to the mission of many Federal agencies. Hundreds of applications in medicine, industry, national security, defense and research depend on isotopes as essential components. Over the years, individual communities and Federal agencies have conducted their own studies, identifying their needs in terms of isotope production and availability. Most recently, the DOE Office of Nuclear Energy and the Office of Science's Office of Nuclear Physics organized a workshop to bring together stakeholders (users and producers) from the different communities and disciplines to discuss the Nation's current and future needs for stable and radioactive isotopes, as well as technical hurdles and viable options for improving the availability of those isotopes.

The next step is to establish the priority of research isotope production and development, and the formation of a strategic plan for the NIPA Program, in which we expect NSACI to play a vital role. The NIPA's products and services are sold world-wide both to researchers and commercial organizations. The NIPA produces isotopes only where there is no U.S. private sector capability or when other production capacity is insufficient to meet U.S. needs. Commercial isotope production is on a full-cost recovery basis. The following two charges are posed to the NSAC subcommittee:

Charge 1

As part of the NIPA Program, the FY 2009 President's Request includes $3,090,000 for the technical development and production of critical isotopes needed by the broad U.S. community for research purposes.

NSACI is requested to consider broad community input regarding how research isotopes are used and to identify compelling research opportunities using isotopes.

The subcommittee's response to this charge should include the identification and prioritization of the research opportunities; identification of the stable and radioactive isotopes that are needed to realize these opportunities, including estimated quantity and purity; technical options for producing each isotope; and the research and development efforts associated with the production of the isotope. Timely recommendations from NSACI will be important in order to initiate this program in FY 2009; for this reason an interim report is requested by January 31, 2009, and a final report by April 1, 2009.

Charge 2

The NIPA Program provides the facilities and capabilities for the production of research and commercial stable and radioactive isotopes, the scientific and technical staff associated with general isotope development and production, and a supply of critical isotopes to address the needs of the Nation. NSACI is requested to conduct a study of the opportunities and priorities for ensuring a robust national program in isotope production and development, and to recommend a long-term strategic plan that will provide a framework for a coordinated implementation of the NIPA Program over the next decade.

The strategic plan should articulate the scope, the current status and impact of the NIPA Program on the isotope needs of the Nation, and scientific and technical challenges of isotope production today in meeting the projected national needs. It should identify and prioritize the most compelling opportunities for the U.S. program to pursue over the next decade, and articulate their impact.

A coordinated national strategy for the use of existing and planned capabilities, both domestic and international, and the rationale and priority for new investments should be articulated under a constant level of effort budget, and then an optimal budget. To be most helpful, the plan should indicate what resources would be required, including construction of new facilities, to sustain a domestic supply of critical isotopes for the United States, and review the impacts and associated priorities if the funding available is at a constant level of effort (FY 2009 President's Request Budget) into the out-years (FY 2009 – FY 2018). Investments in new capabilities dedicated for commercial isotope production should be considered, identified and prioritized, but should be kept separate from the strategic exercises focused on the remainder of the NIPA Program.

An important aspect of the plan should be the consideration of the robustness of current isotope production operations within the NIPA program, in terms of technical capabilities and infrastructure, research and development of production techniques of research and commercial isotopes, support for production of research isotopes, and current levels of scientific and technical staff supported by the NIPA Program. We request that you submit an interim report containing the essential components of NSACI's recommendation to the DOE by April 1, 2009, and followed by a final report by July 31, 2009.

These reports provide an excellent opportunity for the Nuclear Physics program to inform the public about an important new facet of its role in the everyday life of citizens, in addition to the role of performing fundamental research. We appreciate NSAC's willingness to take on this important task, and look forward to receiving these vital reports.

Sincerely,
Jehanne Simon-Gillo
Acting Associate Director of the Office of Science
for Nuclear Physics

APPENDIX 2: MEMBERSHIP OF NSAC ISOTOPES COMMITTEE

Ercan Alp Ph.D.
Argonne National Laboratory
eea@aps.anl.gov

Ani Aprahamian Ph.D. (co-chair)
University of Notre Dame
aapraham@nd.edu

Robert W. Atcher Ph.D.
Los Alamos National Laboratory
ratcher@lanl.gov

Kelly J. Beierschmitt Ph.D.
Oak Ridge National Laboratory
beierschmitt@ornl.gov

Dennis Bier M.D.
Baylor College of Medicine
dbier@bcm.tmc.edu

Roy W. Brown
Council on Radionuclides and Radiopharmaceuticals, Inc
roywbrown@sbcglobal.net

Daniel Decman
Lawerence Livermore National Laboratory
decman1@llnl.gov

Jack Faught
Spectra Gas Inc.
jackf@spectragasses.com

Donald F. Geesaman Ph.D.(co-chair)
Argonne National Laboratory
geesaman@anl.gov

Kenny Jordan
Association of Energy Service Companies
kjordan@aesc.net

Thomas H. Jourdan Ph.D.
University of Central Oklahoma
tjourdan@uco.edu
Steven M. Larson M.D.
Memorial Sloan-Kettering Cancer Center
larsons@mskcc.org

Richard G. Milner Ph.D.
Massachusetts Institute of Technology
milner@mit.edu

Jeffrey P. Norenberg Pharm.D.
University of New Mexico
jpnoren@unm.edu

Eugene J. Peterson Ph.D.
Los Alamos National Laboratory
ejp@lanl.gov

Lee L. Riedinger Ph.D.
University of Tennessee
lrieding@utk.edu

Thomas J. Ruth Ph.D.
TRIUMF
truth@triumf.ca

Robert Tribble Ph.D. (ex-officio)
Texas A&M University
tribble@comp.tamu.edu

Roberto M. Uribe Ph.D.
Kent State University
ruribe@kent.edu

APPENDIX 3: AGENDAS OF MEETINGS I, II, AND III HELD BY NSACI

NSAC Isotopes Subcommittee Meeting I
November 13-14, 2008
Hilton, Gaithersburg, Maryland

Thursday, November 13, 2008

9:00	Welcome
9:15	Charge from NSAC Chair – Robert Tribble
9:30:	DOE-ONP perspective – Jehanne Simon-Gillo
10:00	Introduction – Don Geesaman
10:30	Break
10:45	Overview of the NE Isotopes Program–John Pantaleo
12:00	Lunch
1:30	Report from Isotopes Workshop – John D'Auria
2:15	Discussion of the charge and subcommittee perspective
3:30	Break
3:45	Industry perspective – Roy Brown
4:45	Discussion of the plan forward
5:30	Adjourn

Compelling Research Opportunities using Isotopes 39

Friday, November 14, 2008

9:00   Discussion of how to involve the broad community
10:00   Presentations of recent reports – Tom Ruth: National Academies Study
            Robert Atcher: National Cancer Institute Study
11:30   Executive session
1:00   Adjourn

NSAC Isotopes Subcommittee Meeting II
December 15-16, 2008
Betheseda, Maryland

Monday, December 15, 2008

9:00   Introduction
9:45   OMB – Mike Holland
10:00   FBI – Dean Fetteroff
10:45   Break
11:00   National Institute of Biomedical Imaging and Bioengineering – Belinda Seto
12:00   Lunch
1:30   Department of Homeland Security/DNDO – Jason Shergur
2:10   DOE Office of Nuclear Physics– John D'Auria
2:50   DOE Office of Basic Energy Sciences – Lester Morss
3:30   Break
3:45   National Science Foundation – Brad Keister
4:30   Perspective – Jack Faught 5:00 Perspective – Kenny Jordan
5:30   Adjourn

Tuesday, December 16, 2008

9:00   National Cancer Institute – Craig Reynolds
9:40   NNSA – Victor Gavron
10:30   GNEP – Tony Hill
11:10   Executive session
1:30   Adjourn

NSAC Isotopes Subcommittee Meeting III
January 13-15, 2009
Rockville, Maryland

Tuesday, January 13, 2009

Input from Professional Societies and other groups on priorities for research Speakers and organizations

9:00   Introduction
9:15   Sean O'Kelly, TRTR
10:00   Lynne Fairobent, AAPM
10:40   Break
11:00   Mark Stoyer, ACM/DNCT

11:40	J. David Robertson, MURR
12:30	Lunch
14:00	Gene Peterson, R&D for Accelerator Production of Isotopes
14:40	Scott Aaron, Stable Isotopes
15:30	Break
16:10	Roberto Uribe-Rendon, CIRMS
16:50	Robert Atcher, SNM

Wednesday, January 14, 2009

9:00	Michael Welch
9:40	Richard Toohey, HPS
10:30	Break
11:10-17:00	Executive Session

Thursday, January 15, 2009

9:00-16:00	Executive Session

APPENDIX 4: LIST OF FEDERAL AGENCIES CONTACTED BY NSACI

Air Force Office of Scientific Research
Armed Forces Radiobiology Research Institute
Department of Agriculture
Department of Defense
Department of Energy - Fusion Energy Sciences
Department of Energy- National Nuclear Security Administration - Nuclear Non-proliferation
Department of Energy-Basic Energy Sciences
Department of Energy-Biological and Environmental Research
Department of Energy-Nuclear Physics
Department of Homeland Security
Environmental Protection Agency
Federal Bureau of Investigation
National Cancer Institute
National Institute of Allergy and Infectious Disease
National Institute of Biomedical Imaging and Bioengineering
National Institute of Drug Abuse
National Institute of Environmental Health Science
National Institute of General Medical Science
National Institute of Standards and Technology
National Science Foundation - Directorate for Engineering
National Science Foundation - Directorate for Mathematical and Physical Sciences
National Science Foundation- Directorate for Biological Sciences
Office of Naval Research
State Department
U. S. Geologic Survey

APPENDIX 5: LIST OF PROFESSIONAL SOCIETIES CONTACTED BY NSACI

Academy of Molecular Imaging
Academy of Radiology Imaging
Academy of Radiology Research
American Association of Physicists in Medicine
American Association of Cancer Research
American Chemical Society
American Chemical Society - Division of Nuclear Chemistry and Technology
American College of Nuclear Physicians
American College of Radiology
American Medical Association
American Nuclear Society
American Nuclear Society - Division of Isotopes and Radiation
American Pharmacists Association - Academy of Pharmaceutical Research and Science (APhA-APRS)
American Physical Society
American Physical Society - Division of Biological Physics
American Physical Society - Division of Material Physics
American Physical Society - Division of Nuclear Physics
American Society of Clinical Oncology
American Society of Hematology
American Society of Nuclear Cardiology
American Society of Therapeutic Radiation and Oncology
Council on Ionizing Radiation and Standards
Health Physics Society
National Organization of Test, Research and Training Reactors
Radiation Research Society
Radiation Therapy Oncology Group
Radiochemistry Society
Radiological Society of North America
Society of Molecular Imaging
Society of Nuclear Medicine

In: Isotopes for the Nation's Future
Editor: Ezio Benfante

ISBN: 978-1-61470-818-6
© 2012 Nova Science Publishers, Inc.

Chapter 2

ISOTOPES FOR THE NATION'S FUTURE: A LONG RANGE PLAN[*]

Nuclear Science Advisory Committee Isotopes Subcommittee

Report for the Second of Two 2008
Charges to the Nuclear Science Advisory
Committee on the Isotope Development
and Production for Research and
Applications Program

The Cover: The discovery of isotopes is less than 100 years old. Today we are aware of about 250 stable isotopes of the 90 naturally occurring elements. The number of natural and artificial radioactive isotopes already exceeds 3200, and this number keeps growing every year. "Isotope" originally meant elements that are chemically identical and non-separable by chemical methods. Now isotopes can be separated by a number of methods such as distillation or electromagnetic separation. The strong colors and the small deviations from one to the other indicate the small differences between isotopes that yield their completely different properties in therapy, in nuclear science, and in a broad range of other applications. The surrounding red, white, and blue theme highlights the broad national impact of the DOE Isotope Development and Production for Research and Applications Program.

EXECUTIVE SUMMARY

In 2009, with the signing of the FY09 Omnibus Spending Bill (Public Law 111-8), the Department of Energy's Isotope Production Program was transferred from the Department of Energy (DOE) Office of Nuclear Energy (NE) to the Office of Science's Office of Nuclear

---

[*] This is an edited, reformatted and augmented version of a Nuclear Science Advisory Committee Isotopes Subcommittee publication, dated August 27, 2009.

Physics (ONP). The name of the program has been changed from the National Isotopes Production and Applications Program (NIPA) to the Isotope Development and Production for Research and Applications Program (IDPRA). To prepare for this transfer, the Office of Nuclear Physics and the Office of Nuclear Energy organized a workshop held August 5-7, 2008, in Rockville, MD, that brought together the varied stakeholders in the isotopes enterprise to discuss "the Nation's current and future needs for stable and radioactive isotopes, and options for improving the availability of needed isotopes." The report [NO08] of the "Workshop on the Nation's Needs for Isotopes: Present and Future" is available on the web (http://www.sc.doe.gov/henp/np/program/docs/Workshop%20Report_final.pdf). On August 8, 2008, the DOE-ONP requested the Nuclear Science Advisory Committee (NSAC) to establish a standing committee, the NSAC Isotope (NSACI) subcommittee, to advise the DOE Office of Nuclear Physics on specific questions concerning the isotope program. NSAC received two charges from the DOE Office of Nuclear Physics. The first charge requested NSACI to identify and prioritize the compelling research opportunities using isotopes. NSAC accepted the final report on the first charge in April 2009 and transmitted the report [NS09] to the Department of Energy (http://www. sc.doe.gov/henp/np/nsac/docs/NSAC_ Final_ Report_Charge 1 %20(3).pdf). The second charge is to study the opportunities and priorities for ensuring a robust national program in isotope production and development, and to recommend a long-term strategic plan that will provide a framework for a coordinated implementation of the Isotope Development and Production for Research and Applications Program.

The NSACI subcommittee membership was chosen to have broad representation from the research, industrial, and homeland security communities. During the course of the subcommittee deliberations, a large number of federal institutions, professional societies, industry trade groups, and individual experts were contacted for input.

The mission of DOE's isotope program is threefold:

- Produce and sell radioactive and stable isotopes, associated byproducts, surplus materials, and related isotope services.
- Maintain the infrastructure required to supply isotope products and related services.
- Conduct R&D on new and improved isotope production and processing techniques.

The isotope program is a relatively small federal program (FY08 federal appropriation of $14.8M and FY08 isotope sales of $17.1M) that enables and is immersed in billion dollar enterprises including medical treatment, research, national security, and commercial production and applications. These applications touch the lives of almost every citizen. High priority opportunities are identified in the broad areas of Biology, Medicine and Pharmaceuticals; Physical Sciences and Engineering; and National Security and Applications. Addressing these opportunities effectively will require augmentation of the program's current isotope delivery capabilities.

Numerous reviews of the isotope program have occurred previously. The results of these reviews were considered carefully during the development of this report. The strategic plan developed here, although similar to these previous studies, reflects today's environment and opportunities. The recommendations for the strategic plan are based, in part, on the identification of research opportunities resulting from the first charge.

The production responsibility for certain isotopes does not reside with the isotope program, including available commercially-produced isotopes, isotopes for reactor fuels, and for weapons including plutonium and tritium. Since 2000, the lead responsibility for the conversion of commercial ^{99}Mo (the parent isotope of the most commonly used isotope in medical procedures, ^{99m}Tc) production from processes using highly-enriched uranium to ones utilizing low-enriched uranium has been with DOE/NNSA, in part due to non-proliferation concerns surrounding highly-enriched uranium. Currently, ^{99}Mo is only available from limited foreign sources that have experienced major unplanned supply interruptions over the past few years, leading to serious delays in diagnostic procedures for patients.

The supply of ^{99}Mo, the isotope used to generate the radioactive isotope most frequently used in medical procedures, is of great concern. Recent disruptions in international supply demonstrate the vulnerability of the nation's health care system in this area. The nation must address this vulnerability. At the present time, the isotope program does not produce ^{99}Mo. With the non-proliferation issues associated with the transport and use of the highly-enriched uranium currently used for ^{99}Mo production, DOE/NNSA has been assigned the lead responsibility in this area and is actively investigating options for ^{99}Mo commercial production. The subcommittee chose to refrain at this time from inserting itself into the intense activity underway but reiterates the importance of the issue.

The recommendations of the NSACI subcommittee in response to Charge 2 are divided into three categories: I) Recommendations about the present program, II) Development of a highly skilled workforce for the future, and III) Major investments in production capacity to provide capabilities not available to the nation's current isotope program. The recommendations in the first category are listed in order of priority and the relative priorities of the recommendations in the 2nd and 3rd categories are discussed below.

The Present Program

I.1: Maintain a continuous dialogue with all interested federal agencies and commercial isotope customers to forecast and match realistic isotope demand and achievable production capabilities.

For the isotope program to be efficient and effective for the nation, it is essential that isotope needs be accurately forecast. The DOE-NIH interagency working group is an excellent start for this type of communications in a critical area of isotope production and use.

I.2: Coordinate production capabilities and supporting research to facilitate networking among existing DOE, commercial, and academic facilities.

In the short term, increased isotope production and the availability of new research isotopes require more effectively exploiting the available production facilities including resources outside those managed by the isotope program. This will require both research and development to standardize efficient production target technology and chemistry techniques and flexible funding mechanisms to direct production resources most effectively.

I.3: Support a sustained research program in the base budget to enhance the capabilities of the isotope program in the production and supply of isotopes generated from reactors, accelerators, and separators.

Research and development may significantly expand the production efficiency and capacity of the program. It is also an important path to expanding the skilled isotope production workforce and retaining the most creative people in the program.

I.4: Devise processes for the isotope program to better communicate with users, researchers, customers, students, and the public and to seek advice from experts:

- Initiate a users group to increase communication between isotope program management and users on issues of availability, schedules, priorities, and research.
- Form expert panels as needed to give advice on issues such as definition of isotopes as research or commercial in primary usage, new production methods, and needed actions when demand exceeds supply.
- Modernize the web presence for the isotope program to give users an easier way both to learn about properties, availability, production methods, and services, and also to have access to interactive tools that help customers plan purchases and use, researchers to share information and form collaborations, and students and the general public to learn about the important uses of isotopes.

I.5: Encourage the use of isotopes for research through reliable availability at affordable prices.

Many research applications, and especially medical trials, cannot proceed without a dependable source of isotopes. At the same time, DOE should reexamine its pricing policy for research isotopes to encourage U.S. leadership in isotope-based research.

I.6: Increase the robustness and agility of isotope transportation both nationally and internationally.

- Identify and prioritize transportation needs through establishing a transportation working group.
- Initiate a collaborative effort to develop and resolve the priority issues (i.e., certification of transportation casks).

Highly Trained Workforce for the Future

II: Invest in workforce development in a multipronged approach, reaching out to students, post-doctoral fellows, and faculty through professional training, curriculum development, and meeting/workshop participation.

The dwindling population of skilled workers in areas relating to isotope production and applications is a widely documented concern. This recommendation is focused on the needs of the IDPRA program, itself. The relative priority of this recommendation is comparable to that for a sustained R&D program, with which it is closely linked.

Major Investments in Production Capability

The present program is highly flexible and responsive to the needs of the nation. However, it lacks two major capacities that seriously limit its ability to fulfill its mission. The isotope program presently has no working facilities for the separation of a broad range of stable and long-lived isotopes. Each year it is depleting its unique stockpile of isotopes to the point where some are no longer available. Secondly, many radioactive isotopes by their very nature can be short-lived and cannot be stockpiled. The current program relies on accelerators and reactors whose primary missions are not isotope production; thus, it is not in a position to provide continuous access to many of the isotopes.

III.1: Construct and operate an electromagnetic isotope separator facility for stable and long-lived radioactive isotopes.

It is recommended that such a facility include several separators for a raw feedstock throughput of about 3 00-600 milliAmpere (10-20 mg/hr multiplied by the atomic weight and isotopic abundance of the isotope). This capacity will allow yearly sales stocks to be replaced and provide some capability for additional production of high-priority isotopes.

III.2: Construct and operate a variable-energy, high-current, multi-particle accelerator and supporting facilities that have the primary mission of isotope production.

The most cost-effective option to position the isotope program to ensure the continuous access to many of the radioactive isotopes required is for the program to operate a dedicated accelerator facility. Given the uncertainties in future demand, this facility should be capable of producing the broadest range of interesting isotopes. Based on the research and medical opportunities considered by the subcommittee, a 3 0-40 MeV maximum energy, variable energy, high-current, multi-particle cyclotron seems to be the best choice on which to base such a facility.

The subcommittee gives somewhat higher overall priority to the electromagnetic isotope separator as there is no U.S. replacement. However, a solution in this area is not needed as urgently as the new accelerator capability. Therefore, in the subcommittee's optimum budget scenario that includes both, the construction of the new accelerator starts a year earlier.

The implications of these recommendations are discussed in an optimal budget scenario and under a constant level of effort budget (taken to be the 2009 President's request of $19.9M). Given the recent investments in the isotope program, especially significant American Recovery and Reinvestment Act (ARRA) funding, constant effort funding will allow the program to move forward from a more solid base for a few years. Once this ARRA funding disappears, sustained constant effort level funding, while it does represent a needed increase from the 2004-2008 levels, will place the infrastructure needs for research isotopes at risk in the long term and will not allow the program to address either of the two major missing capacities. The subcommittee does not consider this to be a wise course for the future. The subcommittee recommends an optimum budget that reaches a sustained base operating funding of about $25M (FY09$) per year and also includes new capital funds of about $1 5M (FY09$) per year for several years to realize the needed new capacities.

INTRODUCTION

In 2009, with the signing of the FY09 Omnibus Spending Bill (Public Law 111-8), the Department of Energy's Isotope Production Program was transferred from the Department of Energy (DOE) Office of Nuclear Energy (NE) to the Office of Science's Office of Nuclear

Physics (ONP). The name of the program has been changed from the National Isotopes Production and Applications Program (NIPA) to the Isotope Development and Production for Research and Applications Program. The Office of Nuclear Physics and the Office of Nuclear Energy organized a workshop held August 5-7, 2008, in Rockville, MD, that brought together the varied stakeholders in the isotopes enterprise to discuss "the Nation's current and future needs for stable and radioactive isotopes, and options for improving the availability of needed isotopes." The report [NO08] of the "Workshop on the Nation's Needs for Isotopes: Present and Future" (http://www.sc.doe.gov/henp/np/program/docs/Workshop%20Report_final.pdf) is available on the web. In preparation for the change in program management, the DOE-ONP requested the Nuclear Science Advisory Committee (NSAC) to establish a standing committee, the NSAC Isotope (NSACI) subcommittee, to advise the DOE Office of Nuclear Physics on specific questions concerning the isotope program. On August 8, 2008, NSAC received two charges from the DOE Office of Nuclear Physics. A copy of the full charge letter is attached as Appendix 1.

The first charge requested NSACI to identify and prioritize the compelling research opportunities using isotopes. NSAC accepted the final report on the first charge in April 2009 and transmitted the report to the Department of Energy (http://www. sc.doe.gov/henp/np/ nsac/docs/NSAC_Final_ Report_Charge 1%20(3).pdf) [NS09]. The second charge is

Charge 2:

The NIPA Program provides the facilities and capabilities for the production of research and commercial stable and radioactive isotopes, the scientific and technical staff associated with general isotope development and production, and a supply of critical isotopes to address the needs of the Nation. NSACI is requested to conduct a study of the opportunities and priorities for ensuring a robust national program in isotope production and development, and to recommend a long-term strategic plan that will provide a framework for a coordinated implementation of the NIPA Program over the next decade.

The strategic plan should articulate the scope, the current status and impact of the NIPA Program on the isotope needs of the Nation, and scientific and technical challenges of isotope production today in meeting the projected national needs. It should identify and prioritize the most compelling opportunities for the U.S. program to pursue over the next decade, and articulate their impact. A coordinated national strategy for the use of existing and planned capabilities, both domestic and international, and the rationale and priority for new investments should be articulated under a constant level of effort budget, and then an optimal budget. To be most helpful, the plan should indicate what resources would be required, including construction of new facilities, to sustain a domestic supply of critical isotopes for the United States, and review the impacts and associated priorities if the funding available is at a constant level of effort (FY09 President's Request Budget) into the out-years (FY09-FY18). Investments in new capabilities dedicated for commercial isotope production should be considered, identified and prioritized, but should be kept separate from the strategic exercises focused on the remainder of the NIPA Program.

An important aspect of the plan should be the consideration of the robustness of current isotope production operations within the NIPA program, in terms of technical capabilities and infrastructure, research and development of production techniques of research and commercial isotopes, support for production of research isotopes, and current levels of scientific and technical staff supported by the NIPA Program. We request that you submit an interim report containing the essential components of NSACI's recommendation to the DOE by April 1, 2009, and followed by a final report by July 31, 2009.

The NSACI subcommittee membership was chosen to have broad representation from the research, industrial, and homeland security communities. The membership is given in Appendix 2. Five meetings were called by the subcommittee. The agendas for the meetings are given in Appendix 3. During the course of the subcommittee deliberations, a large number of federal institutions, professional societies, industry trade groups, and individual experts were contacted for input (See Appendices 3-6). Background information is available at the subcommittee web site (http://www.phy.anl.gov/mep/NSACI/).

1) 1995, "Isotopes for Medicine and Life Sciences", Institute of Medicine. [IM95]
2) 1999, "Forecast Future Demand for Medical Isotopes", Expert Panel Review. [NE99]
3) 2000, "Final Report, Nuclear Energy Research Advisory Committee Subcommittee for Isotope Research and Production Planning", Nuclear Energy Research Advisory Committee Subcommittee. [NE00]
4) 2004, "Radiopharmaceutical Development and the Office of Science", Biological and Environmental Research Advisory Committee Subcommittee. [BE04]
5) 2005, "National Radionuclide Production Enhancement Program: Meeting Our Nation's Needs for Radionuclides", Society of Nuclear Medicine National Radionuclide Production Enhancement Task Force. [SN05]
6) 2005, "The U. S. National Isotope Program: Current Status and Strategy for Future Success, American Nuclear Society Special Committee on Isotope Assurance, M. J. Rivard *et al.*, Appl. Rad. Isotopes 63, 157 (2005). [RI05]
7) 2005, "Management of the Department's Isotope Program", DOE Office of Inspector General, Audit Report DOE/IG-0709. [DOE05]
8) 2007, "Advancing Nuclear Medicine Through Innovation, National Research Council Committee on State of the Science of Nuclear Medicine. [NR07]
9) 2008, "Radiation Source Use and Replacement", National Research Council Committee on Radiation Source Use and Replacement. [NR08]
10) 2008, "Report of the Meeting to Discuss Existing and Future Radionuclide Requirements for the National Cancer Institute, Expert Panel Report. [NC08]
11) 2009, "Medical Isotope Production Without Highly Enriched Uranium", National Research Council Committee on Medical Isotope Production Without Highly Enriched Uranium. [NR09]

Figure 1.1. A selection of National Academy and expert panel reports addressing aspects of the isotope program.

There have been numerous expert panels providing advice to the isotope program. Figure 1.1 provides a summary of many of the important reports of the past decade that have a bearing on the program. To many individuals who participated in the work of this subcommittee, from providing input to serving on the committee, there was a sense of deja vu, that the issues and possible courses of action remained very similar. The subcommittee seriously weighed the context and recommendations of each of these previous reports. However, the mission, needs, capabilities, and landscape of competition have evolved, sometimes significantly, over time. The subcommittee made an independent assessment of the current program in today's environment. The recommendations for the long range plan are based, in part, on the identification of research opportunities resulting from the first charge.

For convenience, Appendix 7 reproduces the Tables of priority research opportunities and the recommendations of the first report. These opportunities are further developed in Chapter 3. An interim report containing draft recommendations for a coordinated national strategy and draft budget projections for an optimal budget and a constant level of effort budget was transmitted to NSAC on March 31, 2009. The present document represents the final report for the second charge.

2. HISTORY OF DOE ISOTOPES PROGRAM

The realization of the potential usefulness of isotopes came almost simultaneously with the discovery of the nuclear reactions that could produce them (See Sidebar 2.1 for an explanation of what an isotope is). In 1935, Ernest Lawrence, the inventor of the cyclotron, invited his brother John, a physician, to explore the use of radioisotopes in biology and medicine. In the same year, Rudolf Schoenheimer [SC35A] proposed using stable isotopes for a broad suite of research applications to trace metabolic events in vivo (See Sidebar 5.1). In 1954, when nuclear facilities were not widely available, the Atomic Energy Act directed the Atomic Energy Commission to insure the continued conduct of research and development and training activities in a number of areas including nuclear processes and the utilization of radioactive material for medical, biological, and health purposes. Prices were to be based on an equitable basis to provide reasonable compensation to the government, to not discourage the use of or the development of sources of supply independent of DOE, and to encourage research and development. Under this policy, many extremely valuable uses of isotopes were pioneered for the benefit of the nation. An early application was the use of radioactive iodine to treat thyroid cancer.

In the 1950's, Brookhaven National Laboratory developed the molybdenum-99/technetium-99m generator for the isotope most commonly used in medical procedures today. In the 1970's Fluorine-18 fluorodeoxyglucose (FDG) was developed for positron emission imaging (PET) of cancer. Large quantities of stable isotopes were separated for research purposes. These isotopes became irreplaceable tools for studies of human nutrition (Sidebar 5.1) or for understanding the properties of superconductors and other specialized materials (Sidebar 3.B.2). Many of the Sidebars throughout this report provide other success stories. Radioactive isotope applications became standards for such broad uses as smoke detectors, emergency exit signs, and well logging for oil exploration. As anticipated, with increased use came a substantial commercial market and commercial suppliers. The estimated value of all U.S. isotope shipments in 2007 was about $3 billion [ITS09]. In such cases, issues can arise of the fairness of competition between private and federal sources of a useful product. In general, it is the policy that the federal government does not compete with private industry unless dominant national interests are determined to be involved. The Department of Energy adheres to the procedures and criteria expressed in the Federal Register, Tuesday, March 9, 1965, with respect to determinations involving its withdrawal and re-entry into commercial markets. These include reasonable and consistent prices, but allow a federal position in the market in the case of some single source or foreign producers. Under these procedures, private industry may petition the government to withdraw from a competitive market.

Isotopes for the Nation's Future: A Long Range Plan

Sidebar 2.1: What is an Isotope?

Atoms are composed of a tiny positively charged atomic nucleus, made up of relatively heavy positively charged protons and neutrons, which have no electrical charge, and a cloud of light negatively charged electrons that occupy most of the volume of the atom and characterize how the atom interacts with other atoms. However, the number of protons (Z) in the nucleus determines the number of electrons, and thus the chemical element of the atom. Nature allows nuclei with many possible neutron numbers (N) for the same proton number. These differing arrangements are called the isotopes of an element. While there are only 90 naturally occurring elements, and including man-made elements we know of about 120, there are expected to be about 7000 possible isotopes that live longer that a few nanoseconds. Scientists have only thus far identified about ½ of these.

Some of these isotopes are stable. For example, carbon, which has 6 protons, has two stable isotopes, ^{12}C and ^{13}C (or sometimes denoted C-12 and C-13) where the C identifies the element as carbon and the superscripts 12 and 13 designate the total number of protons and neutrons, and, to an accuracy of about 1%, give the mass of the atom in atomic mass units. Since the isotopes are chemically very similar, these differences in mass are one of the most commonly used ways to separate isotopes, either directly using centrifuges, or by taking advantage of the fact that moving ions with differing charge to mass ratios bend differently in a magnetic field. In nature, about 98.9% of all carbon is ^{12}C and 1.1% is ^{13}C (used in nutrition studies). The difference in abundances is due to substantial differences in the rates of nuclear reactions between isotopes when the elements are created in stars and stellar explosions. In the case of carbon, 11 other isotopes are known with half-lives for nuclear decay between 5715 years for ^{14}C used in radioactive carbon dating, 20 minutes for ^{11}C used in medical diagnosis (positron emission tomography) to 0.009 s for ^{22}C.

While in most cases, only the ground state of a nucleus lives long enough to be useful for applications, there are some instances where an excited state has particularly useful properties. These states are know as "isomers" and are designated with an "m" for metastable. An especially useful isomer occurs in an isotope of technetium with mass 99.

This isomer, ^{99m}Tc, is used in about 15 million medical procedures a year in the United States.

Useful quantities of unstable isotopes typically must be artificially created by man with nuclear reactions using particle accelerators or nuclear reactors (See Chapters 6 and 7). In most cases, stable isotopes can be separated out of naturally occurring materials (See Chapter 5). However if a stable isotope, such as ^{3}He (with an abundance of 0.0001%), is sufficiently rare, it too must be created through man-made nuclear reactions.

In 1990, the Energy and Water Development Appropriations Act (Public Law 101-101) substantially modified the DOE isotope program by requiring "fees shall be set by the Secretary of Energy in such a manner as to provide full cost recovery, including administrative expenses, depreciation of equipment, accrued leave, and probable losses." At the same time, an Isotope Production and Distribution Program Fund was established, and the appropriation and revenues received from the disposition of isotopes and related services were credited to this account to be available for carrying out these purposes without further appropriation. The intent was likely to increase the commercial production of isotopes and to attempt to provide rational, market driven pricing to isotopes. However, the consequences were severe, especially to the research community. For example, in 1998 the primary high-throughput electromagnetic isotope separator facility for the U.S., the Y-12/ORNL Calutrons, was shut down because at full cost recovery it could not compete with foreign suppliers, primarily Russian, who could artificially set lower prices and thus effect a change in the market creating an advantage for their inventory of government produced isotopes. With these fiscal constraints, maintaining the aging facility in a state of readiness was deemed too expensive.

These negative impacts of Public Law 101-101 were appreciated and in 1995, Public Law 103-316 stated "fees set by the Secretary for the sale of isotopes and related services shall hereafter be determined without regard to the provisions of Energy and Water Development Appropriations Act (Public Law 101 -101)." This law, in principle, gives broad latitude to DOE in determining pricing policy. However, each year the President's budget request contains language similar to that in the 2009 request, "The isotope program operates under a revolving fund established by the 1990 Energy and Water Appropriations Act (Public Law 101-101), as modified by Public Law 103-3 16. Each isotope will be priced such that the customer pays the cost of production. The DOE will continue to sell commercial isotopes at full-cost recovery."

Over the course of time, a number of specific missions were added and then removed from the DOE isotope program, in part as new promising applications or treatments or as significant issues in isotope availability were identified. For example, in the 1990's, the isotope program was charged to develop a source for ^{99}Mo (See Sidebar 4.3 for a more complete discussion of the critical issues of ^{99}Mo supply). When Canada committed to developing the new Maple reactors to produce ^{99}Mo, that program in the U.S. was terminated in 1999 due to lack of commercial interest in serving as a domestic source of supply. From 2000 to 2002, an Advanced Nuclear Medicine Initiative provided funding for researchers to develop new isotope technologies and train experts in fields relevant to nuclear medicine. From 2001 to 2002, the program also prepared to process material to obtain ^{229}Th to extract isotopes that had been shown to be effective in treating acute myeloid leukemia in Phase I trials. The program ended with a plan that it would be picked up by the private sector, a hope that has not been realized. The need for increased quantities of these alpha-decaying isotopes remains (See the first recommendation of NSACI Charge 1 report, Appendix 7). The responsibility for managing and disposing of the $_{233}$U material, from which the ^{229}Th would be extracted, was transferred to the DOE-Office of Environmental Management.

Figure 2.1 illustrates the primary facilities currently operated by the DOE that are used the produce isotopes and one notable university facility with a cooperative agreement. The IDPRA program has stewardship responsibilities at three national laboratories. The Brookhaven Linac Isotope Producer (BLIP) uses the linac injector for the DOE-ONP facility, the Relativistic Heavy Ion Collider, to provide up to 200 MeV proton beams of up to 105 μA in both parasitic and dedicated running modes. At Los Alamos, the Isotope Production Facility (IPF) at the Los Alamos Neutron Science Center (LANSCE) provides 100 MeV 400 μA proton beams, again in both parasitic and dedicated running modes. LANSCE's primary support comes from DOE/NNSA. Proton beams of the energies available at BLIP and IPF are not available elsewhere in the United States for isotope production. The 85 MW High Flux Isotope Reactor at Oak Ridge National Laboratory is operated by the DOE-Office of Basic Energy Sciences for neutron scattering research, materials research, and isotope production. The host facilities (RHIC, LANCE and HFIR) are primarily funded to support other missions. Isotope production is a secondary mission. Oak Ridge also houses the Isotope Business Office, Materials Laboratories, and the pool of enriched stable isotopes.

Isotopes for the Nation's Future: A Long Range Plan

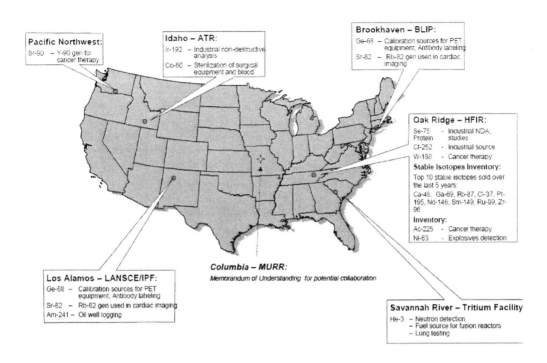

Figure 2.1. Network of DOE production sites. At present, the Isotope Development and Production for Research and Applications Program has stewardship responsibilities for isotope production only at IPF, BLIP, and some of the activities at ORNL.

At each of these three facilities there are also extensive radiochemical laboratories for processing and packaging radioisotopes and the required shipping infrastructure for transporting them safely and efficiently to customers. Also shown in Figure 2.1 is the 250 MW Advanced Test Reactor (ATR) at Idaho National Laboratory operated by the DOE-Naval Reactor program which has been used for ^{60}Co production. Studies are underway to investigate the use of ATR for other isotopes. Pacific Northwest National Laboratory has extensive radiochemical laboratory facilities and is interested in applying this expertise to new chemical processing and applications deployment capabilities. The Savannah River Site and Savannah River National Laboratory provide ^{3}He obtained from the decay of tritium stocks from the dismantlement and maintenance of nuclear weapons. The isotope program coordinates the sales of this ^{3}He.

Special considerations have led to the responsibility for certain isotopes to be assigned to other areas of DOE. These include weapons material such as tritium, enriched uranium, and plutonium. As discussed in Sidebar 4.3, DOE/NNSA has the lead responsibility for ^{99}Mo, in large part due to their non-proliferation responsibilities to reduce or eliminate the use of highly- enriched uranium in the production cycle.

One non-DOE facility is noted in Figure 2.1, the 10 MW Missouri University Research Reactor (MURR). This research reactor facility has a long history and a major program in isotope production, in 2008 making over a 1000 shipments of 49 different radioisotopes to a variety of national and international customers. Recognizing these broad capabilities and the need to ensure multiple isotope production streams, the isotope program has a memorandum of understanding with MURR for potential collaboration.

Beyond the facilities that historically provided a service or are supported by the Isotope Development and Production for Research and Applications Program, there are a large number of university, national laboratory, and commercial accelerators and reactors. These will be discussed in more detail in the following chapters.

In 2008, the appropriated federal budget for the isotope program was $14.8M and revenue from sales totaled $17.1M, for a total budget of $31.9M. Chapter 11 presents the FY09 budget including initiatives funded under the *American Recovery and Reinvestment Act*, and Chapter 12 discusses budget projections for the future under the scenarios specified in the charge.

3. USES OF ISOTOPES

This chapter draws heavily on the material on the uses of isotopes presented in the first report of the NSACI. It gives a glimpse of the priority isotope needs that led to the research priorities presented in the first report (and summarized in Appendix 7) that provide the basis for the consideration of the production capabilities required for the future.

3.A. Biology, Medicine, and Pharmaceuticals

The majority of the isotopic material used in medical and biological research is used to support clinical trials. That said, there is still a significant demand for radioisotopes for use in research during radiochemical, in vitro, and in vivo preclinical investigations. These investigations are critical in the development of radiopharmaceuticals for diagnostic and therapeutic applications.

At the very earliest stage of development, a radionuclide is used to test labeling techniques that are used to radiolabel a targeting molecule. Obviously, given the cost of the starting material, the need to maximize the yield of the reaction is paramount. This is also important in maximizing the specific activity of the compound in question. In many cases, there is a limited number of sites on the target cell. In order to maximize the signal to noise ratio, or to maximize the therapeutic effect, it is important to have the greatest number of radioactive atoms attached to the target. These investigations are often limited to synthesis and analysis using standard techniques such as chromatography.

Once these investigations have been completed, testing of the biologic activity would be undertaken by utilizing cells that express the target that are exposed to the radio-pharmaceutical. Usually, specific binding is determined by utilizing a control cell line of similar characteristics save the target itself and adding a substantial excess of the unlabeled targeting molecule to block specific binding to the target. In the case of a therapeutic conjugate, cell survival assays would also be undertaken to assess the cytotoxicity of the conjugate.

Once a radiopharmaceutical has passed these tests, testing in vivo will be done to assess the compound's ability to target the site of interest. In the case of human tumors, an animal with a compromised immune system will be used so that a xenograft of a tumor of interest can be grown. Usually, administration of the radiopharmaceutical will be via the vein as is the case in the nuclear medicine clinic. In the majority of cases, a combination of imaging of the radiopharmaceutical will be combined with post mortem tissue counting to determine relative and absolute uptake of the material.

Radioactivity can be selectively administered into patients by direct injection and, if the radiation is in the right form, can selectively target human tumors and, if sufficiently concentrated, can destroy the tumors without excessive damage to normal tissues. The prototype for this approach was introduced in the early 1940s, as radioactive iodine, in the form of ^{131}I which could target human thyroid cancer and in some cases, cure patients of metastatic tumors which would have otherwise been fatal. Shortly after this, ^{32}P was introduced for targeting of human bone marrow disorders and was an early and quite effective therapy for abnormal states of myelodysplasia (pre-leukemia disorders) including polycythemia vera (abnormal increase in blood cells, primarily red blood cells, due to excess production of the cells by the bone marrow). These radionuclides were generally introduced in common chemical forms into the body, ^{32}P in phosphate form or ^{131}I as sodium iodide, and through natural processes within proliferating tissues achieve therapeutic concentrations. Although ^{32}P has been supplanted by more selective chemotherapies, ^{131}I continues to be the front-line drug for therapy of advanced thyroid cancer.

The practice of using radioactivity which is introduced into the patient by injection of relatively simple chemical forms continues to this day. For example Holmium and Dysprosium are injected into the joint space of patients for selective therapy of arthritis; ^{153}Sm and ^{89}Sr as simple chelates take advantage of the natural bone seeking properties of this class of chemical element. These two drugs have been approved by the Food and Drug Administration. Bone seeking elements that have slightly improved quality for palliative therapy are also being explored in clinical trials. These include ^{224}Ra and ^{117m}Sn. Both are given as a simple salt, introduced in patients with metastatic prostate cancer, and are effective palliative therapies.

Targeted therapies with radiopeptides or radiolabeled antibodies have been introduced. Patients with non-Hodgkin's lymphoma are now routinely treated, especially in the late stage of their disease, with radioimmunotherapy. This form of modern targeted therapies in medicine takes advantage of knowledge of the biology of cancer and the specific biomolecules that are important in causing or maintaining the neoplastic state (abnormal proliferation of cells). In this case, an antibody or protein is used as the carrier for the radioactivity that confers a specific binding property to a known component of any class of tumors: for example the radiolabeled peptide ^{90}Y (Yttrium DOTATOC) selectively binds to an endocrine receptor on carcinoid tumors, somatostatin type II (growth hormone inhibiting hormone). The targeting occurs much like a key (the radiopeptide) fitting into a lock (the somatostatin receptor), and over time sufficient radioactivity is deposited in the region of the tumor to damage the proliferating capacity of the tumor, in some cases eradicating sites of the tumor completely. In many instances clinical benefit is obtained from the use of the radioactivity, especially in patients with advanced disease. Sidebar 3 .A. 1 describes the successful privatization of ^{90}Y production as a result of the DOE isotope program.

The use of therapeutic radionuclides is expanding in clinical research, and over the course of five years, it is likely that several additional FDA approved clinical applications will become best practice for specific clinical indications.

In general, it is electron/beta or alpha radiation which is most likely to be useful for the purpose of depositing localized radiation in sufficient quantities to kill tumors without damaging normal tissues. Therapeutic radioisotopes are chosen for their radiation properties, including the type of radiation emitted, half-life, and energy. Radionuclides that are proposed for this type of therapeutic radiation usually emit one of three types of radiation: Auger electrons, beta particles, or alpha particles. Sidebar 3.A.2 discusses the properties of each for therapy.

As targeted vehicles including antibodies and peptides become more and more selective for selective binding to biomolecules attached to cancer cells, radionuclides which emit alpha particles have become more and more desirable. These radionuclides have been relatively difficult to get in sufficient quantities. The short-lived alpha emitters are particularly in demand, especially ^{225}Ac, ^{213}Bi, and ^{211}At.

Another area of compelling research with isotopes is in the development of pairs of isotopes in radiopharmaceuticals that can be used simultaneously for therapy and dosimetry. The therapeutic part of this research tests the ability of the radiation to either effectively ablate the tumor as determined by physical measurements or to "cure" the tumor. In order to better gauge the window of effectiveness and toxicity for the therapeutic agent, a surrogate agent is used. The second part of these new developments is the determination of the dosimetry of the compound. This information is then used to determine the dose that would be received by the target tumor and normal tissue without using the therapeutic agent itself. Obviously, the best option would be to use an isotope of the same element so that the chemical issues are the same. Table 3.A.1 presents several examples of such promising therapeutic/dosimetry pairs.

For those compounds that pass these hurdles, patient studies will be undertaken either under the watch of a radioactive drug research committee and the institutional review board or after the investigators have applied to the Food and Drug Administration (FDA) for an Investigational New Drug (IND) status for the compound.

Sidebar 3.A.1: Privatization of ^{90}Y.

Pacific Northwest National Laboratory (PNNL) began developing yttrium-^{90} (90Y) in 1990 in support of the Department of Energy's long-standing mission to create a reliable supply of isotope products, services, and related technologies for use in medicine industry and research. DOE successfully sold ^{90}Y for several years. In 1998 DOE made the decision to privatize the ^{90}Y business. DOE entered into an agreement with New England Nuclear (now Perkin-Elmer) to lease 40 curies of strontium-90, a byproduct of Hanford nuclear weapons production, from the Energy Department over a five-year period, and extract an ultra-pure form of yttrium-90 from the strontium through a process patented at PNNL.

^{90}Y from Perkin-Elmer is currently being used in several clinical trials. Those clinical trials include ^{90}Y DOTA-tyr3-octreotide for the treatment of neuroblastoma, childhood brain tumors, and gastrointestinal cancer, and ^{90}Y edotreotide for the treatment of many types of neuroendocrine tumor cells which are SSTR-positive cells. ^{90}Y from other sources is also being used in Zevalin for the treatment of non-Hodgkin's Lymphoma, and ^{90}Y spheres which are being used for the treatment of liver cancer.

This ^{90}Y privatization by the DOE is a success story pointing out the potential of the national laboratories working with industry for the commercialization of products developed at the labs.

It should be evident that the amount of radionuclide required at each stage of development increases substantially. Thus, in parallel to the biomedical investigations underway, there needs to be a parallel effort to increase the amount of the radionuclide produced to support the research effort.

Table 3.A.1. Pairs of isotopes that can be simultaneously used for dosimetry and therapy

Therapy mechanism	Therapeutic radionuclide	Diagnostic radionuclide for dosimetry	Decay mode of dosimetry agent
Beta decay	^{67}Cu	^{64}Cu	Positron
Beta decay	^{90}Y	^{86}Y	Positron
Beta decay	^{131}I	^{124}I	Positron
Alpha decay (daughter)	^{212}Pb (212Bi)	^{203}Pb	Single Photon

---

Sidebar 3.A.2: Characteristics of Radiation for Therapy.

Electrons, beta particles, and alpha particles are the forms of radiation typically used for targeted therapy because, unlike Y-rays or X-rays, they have relatively short ranges in tissue. It is generally considered that the cell nucleus is the killing zone within a cancer cell. The goal is for the radiation to deposit as much energy as possible within the tumor, while sparing the surrounding normal tissues. Alpha particles are emitted with discrete energies, and, because of their slower velocities and higher charges, deposit a large amount of energy along a relatively short track in tissues. Alpha particles traversing through a cell nucleus will deposit enough energy to kill the cell. Auger electrons are emitted during the process of electron capture decay. Due to their low energies they also deposit energy very densely along the track of their decay.

This characteristic is highly advantageous since if a radionuclide were targeted to a tumor, because then that energy would be deposited within, and maximally damaging the tumor, but sparing surrounding normal tissues. Beta particles (electrons or positrons) are emitted in a decay that involves sharing the total available energy with a simultaneously emitted neutrino, so there is a continuous distribution of electron energies, and the ranges of the electrons at higher energy can be large. Medium energy beta particles (~0.2 MeV) such as those from ^{131}I have a path length of about 400 μm in tissues, while higher energy betas from radionuclides such as ^{90}Y (~2.0 MeV) have path lengths that may range up to 1 cm in tissue. This distribution of energies is undesirable because a significant portion of the deposited energy, especially for small tumors, will be deposited outside the tumor and in normal tissues. Higher energy beta particles are not well suited for treating small tumors.

Identification of the Stable and Radioactive Isotopes that Are Needed to Realize These Opportunities

Stable Isotopes

Virtually all research studies of human *in vivo* metabolism today, in adults as well as children, employ stable rather than radioactive tracers (See Sidebar 5.1). The movement away from radiotracers for such studies came over the last 35 years in large part due to the continued availability of stable isotopes from production programs at Los Alamos and Oak Ridge National Laboratories. The widespread use of ^2H, ^{13}C, and ^{18}O throughout basic and clinical biochemical research has made commercial production of these isotopes feasible, and industry sources are readily available. ^{15}N demand is also met currently by industry sources, but it is not available domestically, and there is no domestic generator of a new inventory. The latter is, potentially, no trivial problem because nitrogen is an indispensible dietary nutrient, especially in its role as the essential nutrient in amino acids, the building blocks of proteins. Thus, since there is no long- lived radiotracer alternative, an absence of ^{15}N would curtail essentially all human studies of nitrogen metabolism. F, Na, Mg, P, S, Cl, K, Ca, Cr, Mn, Fe, Co, Ni, Cu, Zn, Se, Mo and I are essential nutrients in man. Some of these elements (e.g., F, Na, P, Mn, I) are mono-isotopic and, thus, not amenable for use in tracer studies. The remainder, however, have stable nuclides that are critically necessary for investigation of the requirements and metabolism of these indispensible nutrients in humans and animals [FA02, TU06, ST08]. Although these isotopes exist in current DOE inventory, the great bulk of the stable mineral isotopes used for human research are supplied by Russia, and there is great concern for future availability. This concern has been expressed previously [AB92]. It is not an exaggeration to say that research and clinical studies of essential mineral nutrient metabolism in man will come to a complete halt if the supply of these elements is curtailed. These concerns are no less acute or impactful in the domains of studying aquatic and terrestrial ecosystems where, in addition to the nuclides discussed above, the supply of stable isotopes of B, Cd, Ba, Hg, and Pb are, likewise, vitally essential for research into the impact of our environment on biological systems [ST08].

Radioactive Isotopes

The radionuclide and the radiochemical purity of a given isotope of interest are critical. Contaminating radionuclides can degrade the quality of the image, increase the dose to the patient, and render the product unusable according to the specifications for the radiopharmaceutical. Chemical purity matters because the chemical form of the material can potentially reduce the yield of the chemical reactions and potentially reduce the specific activity of the final product if a stable contaminant competes with the radionuclide during synthesis.

Estimated Quantity and Purity of Isotopes of High Priority for Biology, Medicine, and Pharmaceuticals

Research opportunities and priorities were identified in this area in the first report of the NSACI subcommittee and are listed in Appendix 7. The opportunities are listed in priority order for this section. Within each opportunity, if there is particular priority to one isotope, it is noted below. Most of these opportunities followed the recommendations of the "Report of Meeting Held to Discuss Existing and Future Radionuclide Requirements of the National Cancer Institute" [NC08], held on April 30, 2008, the 2007 report of the National Research Council's Committee on the State of Nuclear Medicine, "Advancing Nuclear Medicine Through Innovation," [NR07] and the list of projected isotope needs presented to the Committee by the National Cancer Institute from the on-going DOE-NIH working group.

Alpha therapies have extraordinary research potential, and the isotopes of interest are ^{225}Ac, ^{211}At, ^{213}Bi and ^{212}Pb. Table 3.A.2 gives the quantities of ^{225}Ac or ^{213}Bi that would be needed for various stage clinical trials as reported in the August isotope workshop [NO08]. One important factor for this isotope is that a potentially important interim source of ^{225}Ac is to recover the ^{229}Th parent from stores of ^{233}U that are scheduled to be diluted and disposed of, a process that would make them unsuitable for this purpose. It is estimated that a factor of four more material is needed at this point, and if a Phase II study is undertaken, an order of magnitude more material will be needed. Because rapid action may be needed here, and the linking of ^{225}Ac with another isotope, ^{213}Bi, with the same parent, it is given the highest priority in this opportunity.

Table 3.A.2. Estimated annual usage of ^{225}Ac and/or ^{213}Bi based on known needs. Estimates can vary by .50% depending on whether the approved treatment is with ^{225}Ac or ^{213}Bi [NO08]

Year	Amount(mCi)	Program
2009	1600	Clinical trails (1 multi-center)/R&D support
2010	3100	Clinical trails (2 multi-center)/R&D support
2011	4600	Clinical trails (2 multi-center)/R&D support
2012	7400	Clinical trails (3 multi-center)/R&D support
2013	15000	One approval; Clinical trials (2 multi-center)/ R&D support
2014	50000+	Two approvals; Clinical trials/R&D support

^{211}At is needed in similar amounts, a factor of four more material now and an order of magnitude more should a Phase II study be undertaken. The ^{212}Pb availability is easier to expand than those of ^{225}Ac and ^{211}At since the grandparent ^{232}U has a shorter half-life and can be produced by neutron irradiation of ^{231}Pa.

Several low-energy accelerators located at separate facilities in the United States are currently producing key medical research isotopes (^{64}Cu, ^{124}I). Other medical research isotopes (^{86}Y, ^{203}Pb, ^{76}Br, ^{77}Br) could also be produced at these accelerators. However, since these are research radionuclides and a large commercial market has not been established yet, operators of these accelerators do not have a significant incentive to produce these routinely. As a result, these radionuclides are not always readily available. There are significant advantages foreseen for sharing radiochemistry techniques and targetry technologies across accelerators located around the country in producing these research isotopes. The four diagnostic agents presented in Table 3.A. 1 that are paired with theuraputic agents can all be made at existing low-energy accelerators, but to ensure regular and long term availability, there is a need for increased networking of producers and R&D in order to increase quantities available for researchers. Within this opportunity, priority is not given to any individual pair of isotopes.

A continuously growing need for ^{89}Zr was projected by the DOE-NIH joint working group. This isotope is also produced at lower energy facilities than DOE currently operates and increased and regular availability requires coordination of production and the sharing of production and chemistry techniques. The production of ^{67}Cu requires higher-energy accelerators that are currently available at only three sites, two of which are isotope program facilities. The high demand projected for the future could not be met with current capacities.

An isotope for medical applications should be considered a research isotope until it has been given New Drug Approval (NDA) by the FDA. Preclinical and clinical research subjects are administered these materials under guidance of a radioactive drug research committee or Investigational New Drug status. Until that time, the isotope should be considered as a research material and not subject to petitioning from a private provider. The experience to date has shown that premature abandonment of production has resulted in unsupported increase in price and a spotty ability to meet the demand of the research community to adequately evaluate the effectiveness of a new procedure.

3.B. Physical Sciences and Engineering

Replacing one isotope of an element with another can result in unique responses under various probes in solids, liquids, and gases. This may simply be due to the mass difference of the atomic nucleus, which couples to electronic degrees of freedom; the spin of the nucleus and, therefore, its response to magnetism; or the nuclear structure, which can undergo large variations even with a single neutron addition. This unique behavior allows scientists to directly examine the environment in the sites where the isotopes are added, lending itself to a plethora of useful applications. Thus, in almost all branches of sciences and engineering, from the study of the very small, elementary particles, to the very large, planets and exploding stars, from the study of the very old, geology, to the very new, nanoscience, isotopes have found fundamental and technological applications. For example, isotopes are intimately involved in processes for energy production, industrial diagnostic methods, archeology, geology (terrestrial and extraterrestrial), ecology (carbon and nitrogen cycle), and astronomical science. Isotopes enable the search for new sources of energy, help manage the natural resources like water and forests, and provide for home and food safety.

While the discovery of isotopes is less than 100 years old, today about 250 stable isotopes of the 90 naturally occurring elements are known. The number of natural and artificial radioactive isotopes exceeds 3200 already, and this number keeps growing every year. F. Soddy's discovery [FO 10] in 1910 of lead (Pb) obtained by decay of uranium and thorium differing in mass was considered a peculiarity of radioactive materials. In 1913 Soddy [SO13], and independently Fajan [FA13], developed a displacement law, which explained the change in mass and in the place in the periodic table after .-decay or .-decay takes place and extended its implications on the formation of isotopes.

It is perhaps obvious that isotopes are essential tools in basic research across all of nuclear physics. Indeed, one of the central thrusts expressed in the DOE/NSF Nuclear Science Advisory Committee Long Range Plan for nuclear physics [NS07] is to understand how the properties of the nucleus change as the ratio of the number of neutrons to number of protons varies. This research requires experiments with a variety of isotopic targets and beams. It compellingly leads to the study of ever rarer and rarer isotopes that are far in neutron number from the stable ones. Some of these rare isotopes form the pathway to the formation of many of the elements in the human body during the explosion of a supernova. The Department of Energy plans to construct a major new user facility, the Facility for Rare Isotope Beams at Michigan State University, to provide world leading capabilities for this science. But in the other frontiers of nuclear science, many of the most important experiments depend on reliable and affordable availability of nuclear isotopes. In understanding the nucleon at the fundamental quark and gluon level, targets and beams of ^{2}H and ^{3}He allow access to the neutron. In looking beyond the Standard Model with tests of fundamental symmetries, the important experiments rely on a number of key isotopes.

Specifically, enriched stable isotopes are needed for targets and for accelerated beams at various laboratories producing both stable and radioactive beams needed to study the structure of nuclei. For example, ^{48}Ca is a neutron-rich isotope that is commonly used as a beam at various nuclear physics laboratories to study the properties of exotic nuclei far from stability. Also, it is used in fragmentation reactions to produce very exotic radioactive beams. A future supply of stable highly-enriched isotopes of many different elements is necessary for forefront experiments in nuclear physics. Scientists are also creating new elements in the periodic table and establishing their unique chemical attributes. These latter experiments require actinide targets, including various isotopes of uranium, neptunium, plutonium, americium, curium, californium and berkelium. Research in actinide chemistry also is important for environmental studies of the migration of plutonium and other actinides and the effective disposal of nuclear waste.

The Argonne Tandem Linac Accelerator System (ATLAS) is a DOE-funded national user facility for the investigation of the structure and reactions of atomic nuclei in the vicinity of the Coulomb barrier. A major advance in rare-isotope capabilities at ATLAS will be the Californium Rare Ion Breeder Upgrade (CARIBU). Rare isotopes will be obtained from a one- Curie ^{252}Cf (Californium) fission source located in a large gas catcher from which they will be extracted and accelerated in ATLAS. CARIBU will provide accelerated neutron-rich beams with intensities up to 7×10^5 particles/s, and will offer unique capabilities for a few hundred isotopes, many of which cannot be extracted readily from existing Isotope Separator On Line (ISOL) type sources. In addition, it will make these accelerated beams available at energies up to 10-12 MeV/nucleon, which are difficult to reach at other facilities. As discussed in Sidebar 4.2, the availability of ^{252}Cf from the isotope program for this purpose has been in question. Without continued availability of these ^{252}Cf sources at about 1 Ci every 1.5-2 years, CARIBU cannot fulfill its scientific promise.

A very powerful probe of physics beyond the Standard Model of particles and interactions is to search for a *permanent electric dipole moment (EDM)* of a quantum system. The principles of quantum mechanics tell us that the interaction between an EDM and an applied electric field E is proportional to $S \cdot E$, where S is the spin of the object. This interaction energy changes sign if time is reversed (labeled as a T transformation). In the Standard Model, the predicted effects that violate time reversal invariance are very weak. Indeed, the very fact that the observable universe is made of matter and not an approximately equal mix of matter and anti-matter is a compelling signal that time reversal must be violated at a much larger level than the Standard Model allows. Searches for permanent electric dipole moments are one of the most sensitive probes for this new physics. But these experiments require special isotopes. In the search for an electric dipole moment of the neutron, ^{3}He is required to align the spin of the neutrons and precisely determine the magnetic environment. Certain radioactive atoms possessing a large octupole deformation are expected to have greatly enhanced sensitivity to time-reversal violating forces in the nucleus (Sidebar 3.B.1). Both ^{225}Ra and ^{223}Rn show promise as potential high-sensitivity deformed nuclei. Currently, experiments using these nuclei are being planned or pursued at laboratories around the world, including Argonne National Laboratory (using ^{225}Ra extracted from a ^{229}Th source at ORNL) and TRIUMF in Canada (using a radioactive beam). The precision of the ^{225}Ra experiment is projected to be limited by the current isotope supply.

Sidebar 3.B.1: Search for Violations of Time Reversal Symmetry: A Test of the Standard Model of Particle Physics

An experimental proof of the existence of a permanent electric dipole moment (EDM) of an elementary particle would indicate a violation of time-reversal symmetry, and it would require a modification to the Standard Model, the currently accepted description of elementary particles and the interactions between them. The size of a possible EDM is expected to be larger in a few heavy radioactive nuclei with unusual pear-shaped deformations, like ^{225}Ra (radium-225).

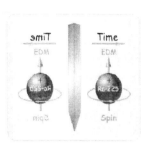

Enhancement in ^{225}Ra from octupole deformation

Figure 1. Under T-reversal, the spin direction reverses while the EDM direction remains the same, thus a particle that possesses both an EDM and a spin is converted into a different kind of particle, and T-symmetry is violated.

	^{199}Hg	^{225}Ra
I	1/2	1/2
$t_{1/2}$	Stable	14.9 d
d ($10^{25}\eta$ e cm)	5.6	2100

With no correlation to spin:

$$\langle \Psi^+ | d_{int} | \Psi^+ \rangle = 0$$

But, with a T-, P-odd interaction V_{PT}:

$$\Psi = \Psi^+ + \alpha\Psi^-$$

$$\alpha = \frac{\langle \Psi^+ | V_{PT} | \Psi^- \rangle}{\Delta E}$$

So, in the lab frame we see:

$$\langle d_z \rangle = 2\alpha d_{int} \frac{I}{I+1}$$

^{229}Th $\xrightarrow{\alpha}$ ^{225}Ra $\xrightarrow{\beta}$ ^{225}Ac $\xrightarrow{\alpha}$ Fr, At, Rn... $\xrightarrow{\alpha,\beta}$ ^{209}Bi
7300 yr — 15 days — 10 days — ~ 4 hours — stable

Currently, the most stringent limits on *T*-reversal symmetry violating interactions in the nucleus are set by experiments that determined a limit on the atomic EDM of ^{199}Hg ($< 3 \times 10^{-29}$ e-cm). The sensitivity to these effects in ^{225}Ra is expected to be 2-3 orders of magnitude larger than in ^{199}Hg. A measurement on ^{225}Ra atoms cooled and confined by laser light in an optical dipole trap offers a promising path. The estimated need for ^{225}Ra is a regular supply of about 200 ng or 10mCi every two months for the experiment. ^{225}Ra is obtained from the ^{229}Th decay chain, which illustrates that the need for ^{225}Ra competes with the need for the isotopes with promising medical applications, ^{225}Ac and ^{213}Bi.

Neutrinoless double beta (0νββ) decay experiments could determine whether the neutrino is its own antiparticle, and, therefore, whether nature violates the conservation of total lepton number. Violation of this symmetry of the Standard Model is another path to the key to the predominance of matter over antimatter. Multiple 0νββ experiments using different isotopes and experimental techniques are important, not only to provide the required independent confirmation of any reported discovery but also because different isotopes have different sensitivities to potential underlying lepton-number-violating interactions.

CUORE - *the Cryogenic Underground Observatory for Rare Events* - is a bolometric detector searching for 0νββ in ^{130}Te. The Italian–Spanish–U.S. collaboration plans to install and operate TeO2 crystals containing 200 kg of ^{130}Te at the underground Laboratori Nazionali del Gran Sasso in Italy. Replacing the natural Te with isotopically enriched material in the same apparatus would subsequently lead to a detector approaching the ton scale.

The Majorana collaboration is engaged in a research and development effort to demonstrate the feasibility of using hyperpure germanium (Ge) diode detectors in a potential one-ton-scale $0\nu\beta\beta$ experiment. The initial Majorana research and development effort, known as the Majorana Demonstrator, utilizes 60 kg of Ge detectors, with at least 30 kg of 86% enriched ^{76}Ge in ultra- low background copper cryostats, a previously demonstrated technology. This Canadian– Japanese–Russian–U.S. collaboration is in close cooperation with the European GERDA Collaboration, which proposes a novel technique of operating Ge diodes immersed in liquid argon. Once the low backgrounds and the feasibility of scaling up the detectors have been demonstrated, the collaborations would unite to pursue an optimized one-ton-scale experiment.

Several other promising opportunities to carry out sensitive $0\nu\beta\beta$ experiments exist, and U.S. nuclear physicists have indicated an interest in being involved. One notable experiment is known as SNO+, a proposed ^{150}Nd-doped scintillator measurement that would utilize the previous Canadian Sudbury Neutrino Observatory hardware of the acrylic sphere, photomultiplier tubes, and support systems in a coordinated international program of $0\nu\beta\beta$ measurements.

The large scales of the isotope requirements of these double-beta experiments are extraordinary. In addition, the samples must be extremely radiologically pure and likely would require underground detector construction to limit cosmic-ray activation.

An isotope that is broadly used in nuclear physics as well as low temperature physics is ^{3}He. As discussed in Chapter 3.C, ^{3}He is also widely used as a neutron detector both for research and engineering and national security needs. Polarized ^{3}He is used as an effective polarized neutron in scattering experiments, e.g., at Jefferson Lab. There are plans to implement a polarized ^{3}He source at BNL to provide polarized neutron beams at the Relativistic Heavy Ion Collider (RHIC). As discussed above, ^{3}He is also a central element in the neutron EDM experiment planned for the SNS.

Many unusual phases of matter like superfluidity, superconductivity, and Bose-Einstein condensation occur at extremely low temperatures, which enable the study of subtle behaviors that are obscured by thermal motion at higher temperature. To reach a temperature below 0.3 K, a key technology is the ^{3}He-^{4}He dilution refrigerator because it can operate continuously, provide a substantial cooling power at temperatures from around 1.0 K down to 0.010 K and below, and can run uninterrupted for months. The ^{3}He-^{4}He dilution refrigerator is also required for experiments that require temperatures as low as 0.00 1 K because it can be used to pre-cool the adiabatic demagnetization systems.

Mass differences between different isotopes cause sufficient change in bond strength and the vibrational characteristics of volatile compounds of H, C, N, and O to affect their heat of vaporization. Thus, time, temperature, and geographical variations of isotope ratio differences can be used as a tracer of climate change and help quantify the hydrogen, carbon, nitrogen, and oxygen cycle on earth. Sources of isotopes are essential as calibration standards. For example, in *Paleoclimatology*, which studies climate change over the entire history of the Earth, oxygen isotope ratios [NA93] play an important role. Water with oxygen-16, $H_2^{16}O$, evaporates at a slightly faster rate than $H_2^{18}O$; this disparity increases at lower temperatures. The $^{18}O/^{16}O$ ratio provides a record of ancient water temperature. The measured heat capacity difference between $H_2^{18}O$ and $H_2^{16}O$ is 0.83 ± 0.12 J K^{-1} moi^{1} for liquid water [NA93]. When global temperatures are lower, snow and rain from the evaporated water tends to be higher in ^{16}O, and the seawater left behind tends to be higher in ^{18}O. Marine organisms would then incorporate more ^{18}O into their skeletons and shells than they would in warmer climates. Paleoclimatologists directly measure this ratio in the water molecules of ice cores or the limestone deposited from the calcite shells of microorganisms. Calcite, $CaCO_3$, takes two of its oxygens from CO_2, and the other from the seawater. The isotope ratio in the calcite found in the skeletons and shells of marine organisms is, therefore, the same as the ratio in the water from which the microorganisms of a given layer extracted, after readjusting for CO_2.

Nitrogen isotopic ratios also provide a powerful tool for evaluating processes within the nitrogen cycle and for reconstructing changes in the cycling of nitrogen through time. The biologically- mediated reduction reactions that convert nitrogen from nitrate (NO_3 $^{-1}$, +5 oxidation state) to nitrite (NO_2 $^{-1}$, +3) to nitrous oxide (N_2O $^{+1}$), to nitrogen gas (N_2 ^{0}), and to ammonia (NH_3 $^{-3}$) are faster for ^{14}N than for ^{15}N as a result of the higher vibrational frequency of bonding to ^{14}N than to ^{15}N. This results in products that are ^{15}N-depleted relative to the substrate. If the substrate reservoir is either closed or has inputs and outputs that are slow relative to one of the reduction processes then the reservoir will become enriched in ^{15}N. Therefore, the stable isotope ratio of nitrogen can be a promising proxy for delineating the *eutrophication* in the environment, which is a process describing an increase in chemical nutrients — compounds containing nitrogen or phosphorus — in an ecosystem. Since nitrogen is one of the important nutrient elements in lakes and abundant in anthropogenic sewage and chemical fertilizers, the range in fractionations of nitrogen isotope ratios in aquatic processes makes nitrogen isotope ratios an excellent tracer to monitor eutrophication.

In *astrophysics and planetary sciences*, measurements of D/H, $^{13}C/^{12}C$, $^{15}N/^{14}N$, or $^{18}O/^{16}O$ of primitive solar system materials record evidence of chemical and physical processes involved in the formation of planetary bodies and provide a link to materials and processes in the molecular cloud that predated our solar system. Modern developments exploiting secondary-ion-massspectroscopy (nano-SIMS) methods have provided mineralogical and isotopic evidence of origins of stardust as composed of precursors of the solar system [MC06]. In all these isotopic ratio techniques, from paleoclimatology to planetary science, the isotope production requirements are for measurement standards.

In *solid-state physics*, vibrational spectroscopy methods, such as Brillouin light scattering or Raman spectroscopy, play a major role in using "isotope labeling," in applications such as identifying the origins of meteorites, or magnitude of atomic displacements in a complex molecule. In superconductivity, the shift in transition temperatures with isotopic substitution is a well-established approach to understand the mechanisms of formation of Cooper pairs and their physical location inside complex crystals. The presence of mixed isotopes also acts as scattering centers in an otherwise perfect crystal, reducing cooperative behavior of atoms with substantially reduced thermal conductivity. Nuclei with unpaired spins can couple with electron spins, and the longer relaxation time of the nuclear spin offers potential as a solid-state quantum memory. Isotopically enriched silicon or germanium-based semiconductors lend themselves for engineered nanostructures with phase coherence quality suitable for solid-state quantum memory devices. In chemistry, elusive transition states in reaction chemistry can be revealed through isotopic labeling. Exploiting the variations in nuclear energy levels between different isotopes leads to isotope-based spectroscopic methods, such as *Mössbauer spectroscopy*, which is a major research tool across many scientific disciplines. For example, decay of ^{57}Co, through an electron capture process to ^{57}Fe, provides an ideal parent/daughter relationship that lends itself to study in hyperfine interactions in magnetism, lattice dynamics, and local atomic structure in condensed matter in an unprecedented energy resolution of 10^{-13} or better. Over 50,000 papers have been published in Mössbauer spectroscopy, and a total of 114 isotopes have been used. Today many of the parent/daughter isotopes are available only from Russia, which is a cause for concern for the scientific community. Mössbauer isotopes are produced either in a cyclotron via deuterium bombardment or in a reactor. A modern application of Mössbauer spectroscopy is discussed in Sidebar 3.B.2.

In determining *fundamental constants and metrology*, developing a mass standard in fundamental units has been a struggle. The current approach, dubbed Avogadro's project, is an ongoing international collaboration between laboratories in Germany, Italy, Belgium, Japan, Australia, and USA to redefine the kilogram in terms of the Avogadro constant. The Avogadro constant is obtained from the ratio of the molar mass to the mass of an atom, and it is known to an uncertainty of 0.1 ppm. The goal is to reduce this to 0.01 ppm by measuring the volume and mass of isotopically pure silicon spheres. For a crystalline structure such as silicon, the atomic volume is obtained from the lattice parameter and the number of atoms per unit cell. The atomic mass is then the product of the volume and density. The limiting factors are the variability from sample to sample of the isotopic abundances of Si and the content of impurities and vacancies. Thus, kilograms of isotopically pure ^{28}Si are needed. Currently two such 1 kg spheres are available. The new spheres were made from just one isotope: ^{28}Si. The mono-isotopic silicon was made in Russia while the near perfect crystal was grown in Germany, and perfect spheres were cut in Australia. To achieve the required concentration of the ^{28}Si isotope, a new centrifugal method was used for producing stable isotopes. SiF_4 of natural isotopic composition was used as a compound for centrifugal enrichment of ^{28}Si. A special centrifugal setup and a technology for production of $^{28}SiF_4$ with extremely high concentrations were developed in the Tsentrotekh-EKhZ Science and Technology Center. As a result, $^{28}SiF_4$ with an isotopic purity of 99.992–99.996% was produced.

A very practical but important power-source type application is *radioisotope thermoelectric generators (RTG)* (See Sidebar 3.B.3). For example, they have been used as power sources for spacecrafts (*Apollo, Pioneer, Viking, Voyager, Galileo, Cassini*), where a few hundred watts of power is needed for a very long time. They can also be used in very practical and large-scale applications like driving pacemakers and other implanted medical devices, where microwatt s of power are needed. Various technologies are under development including Stirling heat engines (devices that convert heat energy into mechanical power by alternately compressing and expanding a fixed quantity of air or other gas, the *working fluid,* at different temperatures) and thermo-photovoltaic devices using piezoelectric materials combined with MEMS (micro-electromechanical systems) technology. The most suitable isotope for RTG applications is ^{238}Pu. It is an alpha emitter; thus it has the lowest shielding requirements and long half-life (87.7 years) high density (19.6 g/cc) and reasonably high energy density (0.56 W/g). While there are concerns for environmental and other safety concerns, potential improvements in energy efficiency and prevention of radiation damage for some piezoelectric converters may increase the electrical conversion efficiency by a factor 10 or more, thus making RTGs even more attractive power sources and, in some cases, perhaps the only alternative. Therefore, the need for alpha emitting isotopes of ^{238}Pu, ^{244}Cm, and ^{241}Am, and beta-decaying ^{90}Sr will continue in the future [KO06].

Sidebar 3.B.2: Mössbauer Isotopes in the Synchrotron Era

More than half of the elements in the periodic table have Mössbauer active nuclei. The Mössbauer effect is related to recoilless absorption and emission of gamma-rays from nuclei bound in a solid. In order to conduct the experiments, however, there is also a need for a suitable parent isotope. For example, the most common Mossbauer probe of all times, ^{57}Fe, needs a parent, ^{57}Co, to decay via electron capture to populate the 5/2 isomeric state of ^{57}Fe, which, in turn, cascades down to the 3/2 spin state at 14.4 keV above ground state, and finally, to the ground state.

Since its discovery in 1957, which resulted in the award of the 1961 Nobel Prize to its discoverer, Dr. Rudolph Mössbauer, there are over 55,000 scientific refereed papers reported. Since 1985, it has become possible to use synchrotron radiation as a radiation source instead of a radioactive isotope, and since 1995, it has become possible to record the phonon density of states of materials containing one or more Mössbauer isotopes.

In a recent study the partial phonon density of states of all elements in a technologically important material, namely, a skutterudite compound of $EuFe_4Sb_{12}$ has been measured. This material has the much sought-after "phonon glass-electron crystal" quality that increases the figure of merit in thermal-to-electric heat conversion efficiency for the RTG generators discussed in Sidebar 3.B.3. Here, all three elements have suitable Mossbauer isotopes: ^{151}Eu, ^{57}Fe, and ^{121}Sb. The crystal structure of this compound is shown in the figure below, showing the "rattling" purple Eu atoms in yellow Sb cage.

As a result of this nuclear resonant inelastic x-ray scattering study using all three isotopic resonances of Eu, Sb, and Fe, it has been demonstrated that Eu atoms in the cage have an uncoupled-mode "rattling" vibrational mode at 7 meV (green curve in the figure below). The Mössbauer isotopes that are exploited in such studies include ^{57}Fe, ^{83}Kr, ^{119}Sn, ^{121}Sb, ^{125}Te, ^{149}Sm, ^{151}Eu, ^{161}Dy, and more.

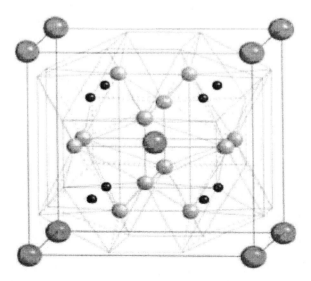

Figure. The crystal structure of $EuFe_4Sb_{12}$. The Eu is purple, Fe is blue, and Sb is yellow.

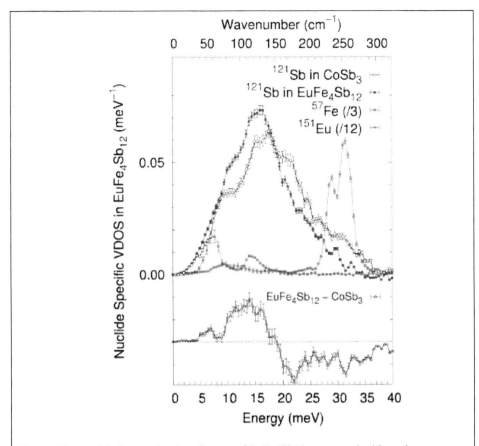

Figure. The partial phonon density of states of EuFe4Sb12, measured with nuclear resonant inelastic x-ray scattering, using the ESRF (Grenoble) and the APS (Argonne).

The stable lithium isotopes, ^{6}Li, and ^{7}Li, have long been used in a number of extremely important research and engineering applications due to their special nuclear and chemical properties and low density. ^{6}Li is particularly important for thermonuclear applications such as nuclear weapons, targets for tritium production and fusion reactors but is also used in advanced battery research. ^{7}Li is currently used primarily for research and for pH balance in boiling and pressurized water nuclear reactors. Typical isotope sales over the past five years have been about 20 kg of ^{6}Li and 4 kg of ^{7}Li per year. The primary technique previously used for lithium isotopic separation was a mercury amalgam process, with significant environmental and human health concerns. As a result, now most of the U.S. production of ^{6}Li is obtained from reprocessing material in the dismantlement of nuclear weapons.

There are several future applications that would result in a major increase in lithium utilization, far beyond current quantities. Advanced fusion power systems could require 10000-40000 kg of ^{6}Li per application. NASA is also considering ^{6}Li as the light-weight shielding of choice for future space based reactors. Typically 1000 kg of ^{6}Li would be needed per reactor. Lithium is also used as the working fluid in a number of advanced nuclear reactor concepts, such as the Advanced High Temperature reactor. In these cases, separated ^{7}Li is required to minimize tritium production. The requirement for a 1 GWe commercial reactor is estimated to be about 25000 kg of 99.995% enriched ^{7}Li. New processes need to be developed and proven to address such large-quantity and high-enrichment needs in an environmentally responsible fashion.

> ### Sidebar 3.B.3: Radioisotope Thermal Generators to Explore the Planets
>
> Isotopes like ^{90}Sr, ^{238}Pu, and ^{244}Cm have been used in radioisotope thermal generators (RTG) to provide power for remote applications like spacecrafts, weather and tsunami warning stations, navigation signal stations, radio beacons, undersea deep-water installations, and transmitters at remote locations with hostile environmental conditions. The power output can be as high as 2.5 W/g and 26 W/cc for ^{244}Cm, making kilowatt power outputs practical. RTG's have been used successfully on 23 spacecrafts since 1961, including planetary (Pioneer, Voyager, Galileo, Ulysses, Cassini, New Horizons), Earth orbit (Transit, Nimbus, LES), lunar surface (Apollo ALSEP), and Mars surface (Viking) probes.
>
> While ^{90}Sr and ^{238}Pu work for this application, their lower power makes the launch masses high. It should be noted that ^{90}Sr is the cheapest isotope to acquire. ^{244}Cm is, perhaps, the best choice because of its high power. It can be recovered through proven methods in large quantities in spent nuclear reactor fuel. Several issues remain for ^{106}Ru, ^{144}Ce, ^{210}Po, or ^{242}Cm, such as difficulties associated with the fabrication of hot isotopes and higher shielding requirements.
>
> Continued discoveries in thermoelectric materials like transition metal antimonides, skutterudites, PbTe, and SiGe, combined with computer aided design of layered systems, provide a promising prospect for RTG's.
>
>

This section has highlighted a few of the many uses of stable and radioactive isotopes in the physical sciences and engineering. Table 9 of the first NSACI report (reproduced in Appendix 7) lists the identified research opportunities where a shortage or potential shortage of isotope supply is an issue, ordered by priority in the physical sciences and engineering areas. Within each opportunity, no particular ordering of priority for individual isotopes has been assigned when more than one isotope is mentioned. The prioritizations are based on the subcommittees' expertise and the priorities presented to NSACI from the DOE-ONP and DOEBES programs. The lighter tone of blue in Table 9 highlights the relatively higher research opportunity potential of these topics.

3. C. National Security and Other Applications

Isotopes are used in many areas related to nuclear security. DHS, NNSA, and the FBI require radioisotopes for the calibration and testing of instrumentation used for the analysis of nuclear materials. NNSA also performs nuclear physics measurements that utilize radioisotopes for calibration and testing purposes. In addition, these organizations use enriched stable isotopes for calibration and isotope dilution measurements in mass spectroscopy. All of these activities require relatively small amounts of these materials, and recently, there have been no major difficulties in supplying these needs. However, in addition to these operational needs, there are some other activities that require larger quantities of materials or materials that are more difficult to obtain.

The U.S. Department of Homeland Security (DHS) is currently deploying many large radiation detection systems to monitor cargo that enters the United States. These devices measure both gamma-rays and neutrons and use this information to detect the presence of nuclear materials. The neutron detectors in these devices use ^{3}He tubes; this type of detector has excellent stability and high efficiency for detecting neutron radiation from plutonium. DHS plans on deploying a large number of these detectors in the course of the next five years. In a similar manner, the "Second Line of Defense" program in NNSA/NA-25 will also deploy a large number of these same types of detection systems in foreign ports that ship cargo to the U.S. The Department of Defense also requires a significant amount of ^{3}He for their portal monitors. An estimate for the combined requirement for ^{3}He by DOE, DHS and DOD for FY09-FY14 alone is greater than 150 kliter. However, the projected supply available from the NNSA stock at Savannah River for FY09-FY14 is only about half this value. The subcommittee is encouraged that DOE is participating in a working group with other government agencies to address this issue.

The U.S. domestic safeguards program uses a combination of destructive and nondestructive analyses to help keep track of special nuclear materials. These techniques require the use of radioactive sources as calibration materials and as sources of active interrogation radiation. In general, the mass measurements of special nuclear materials rely on neutron counting or calorimetry. The high density and high photon attenuation properties of plutonium and uranium limit the effectiveness of gamma-ray measurements. However, gamma-ray measurements of waste contaminated with plutonium or uranium can give an assay of the mass of these materials. One device that performs such measurements is the segmented gamma scanner. This instrument uses high resolution gamma-ray spectroscopy to identify the isotopes of interest, for example, ^{239}Pu, by one or more characteristic gamma rays. A collimator limits the field of view of the detector such that only a segment of the container is visible. The system scans a container of material, and each segment is analyzed for the gamma-ray flux for the isotope of interest. An additional scan of the container is performed with a transmission source, such as ^{75}Se, to determine the attenuation characteristics of the materials in the container as a function of gamma-ray energy. The transmission source is chosen as a radioisotope with a reasonably long half life, >100 days, and with gamma-ray emission lines close in energy to those of the isotope of interest. The transmission corrected count rates of the isotope gamma rays are then used to produce an assay for each segment.

The mass of bulk plutonium samples can be determined by counting the correlated neutrons from the fissioning isotopes with large arrays of ^{3}He tubes and applying coincidence techniques. These multiplicity counting techniques analyze the relative count rates of single neutrons detected in a specific time window versus double or triple detections of neutrons in the same time interval. One can determine the mass of plutonium, the neutron multiplication factor for the sample, and the uncorrelated neutron singles rate (from events such as (a,n) reactions on light elements) from these data. These instruments take advantage of the unique characteristics of the neutron spectrum of ^{252}Cf sources for their efficiency calibration.

The mass of uranium can be determined by counting delayed neutrons from the fission fragments produced by irradiation of a neutron source. One such device, the Californium Shuffler, uses a rapidly moving source of ^{252}Cf to produce a series of short irradiation times for the sample. An analysis is then performed at a time several seconds after the Cf source is removed to determine the amount of uranium present.

Part of the U.S. government's program on nuclear nonproliferation deals with the disposition of excess fissile material from dismantled nuclear weapons. These activities involve the creation of bilateral agreements with the Russian Federation (RF). The agreements can include nondestructive assay techniques to characterize the types of materials, i.e. weapons grade plutonium or highly enriched uranium, and to measure the mass of these materials. These measurements involve sensitive data. Because of the need to protect such information, special measurement methods are being developed to perform these verifications without the disclosure of certain information. These techniques use "information barrier" technology. Radioisotopes play a key role in the development and implementation of these techniques. Also, certified standards of radioactive materials are needed as authentication sources to independently verify that these instruments are producing the appropriate results.

The radioactive isotopes ^{109}Cd, ^{241}Am, and ^{57}Co have been successfully used as transmission sources in systems that monitor the enrichment of uranium as UF6 flowing in the pipes of blending systems and enrichment plants. These systems use the attenuation of photons from these sources (Ag x-rays, 59.5 keV, or 122 keV) to determine the uranium density of the gas in the pipe. The enrichment can be determined by combining the density information with the count rate of the 185.7-keV gamma ray from the decay of ^{235}U.

The isotope ^{252}Cf has also been used in nonproliferation studies as a calibration source for plutonium mass assay equipment and as an interrogation source. An example of the latter is the equipment developed for the U.S. government's transparency program associated with the US/RF highly enriched uranium (HEU) purchase agreement. This device uses a ^{252}Cf source to determine the fissile mass flow of UF6 in the blenddown of HEU to low-enriched uranium.

The area of weapons physics also requires the use of isotopes. With the cessation of nuclear testing, the challenge for the national nuclear security program has been to certify the safety and reliability of the enduring stockpile. Central to this was the realization that the "parametric" engineering-based development program that historically served the program well would have to be modified to have increased emphasis on a more fundamental scientific understanding of weapon performance. With the development of the DOE Accelerated Strategic Computing Initiative based supercomputer capabilities it has become possible to computationally investigate the evolution of a nuclear explosion at an unprecedented level. However, this procedure will only result in a reliable predictive capability if commensurate effort is expended to insure that correct underlying physical data is used in the codes.

The nuclear processes occurring in the explosion are the fundamental heart of the device. A correct understanding of the nuclear reactions and their resultant radiation and particle transport must be accomplished. To this end, nearly every test has utilized "radchem" detectors to provide spatially resolved information on the device performance. The archived data from these tests represents a treasure of detailed information that can provide improved understanding of the underlying weapons physics. These radchem detectors often have been used to diagnose the 14 MeV neutrons produced in the thermonuclear reactions. In the high neutron fluence environment of a nuclear device, multiple nuclear reactions can occur on single radchem detector atoms. These higher order reactions often occur on radioactive isotopes for which little, if any, experimental data exist for their reaction cross sections. Since the radchem production is analyzed at times long compared with the explosion process, these materials are exposed to the complete integral fluence of the produced neutrons. In particular, as the neutrons evolve during the explosion dynamics, they are down-scattered in energy eventually approaching some local environmental thermodynamic equilibrium. At these lower energies the dominant reaction becomes neutron capture. These "late time" effects can result in a perturbation of the isotopic abundances produced in the early thermonuclear burning of the device.

Sidebar 3.C.1: ^{137}Cs as a Calibration Standard for Health Physics

Of the 1430 sources identified as potential candidates for the calibration of radiation detection instrumentation, a relatively few are used to sustain an inter-dependent nationwide calibration network between national standards and secondary standards laboratories. Cs-137 is unique in that it provides mono-energetic photons with an energy of 662 keV; this energy is at the center of the range typically needed for characterizing radiation detection instrumentation and upon which calibrations can be based. ^{137}Cs also has low dose-rates that approximate radiation exposures to which humans might be exposed.

^{137}Cs sources are used to calibrate personnel dosimeters that monitor the radiation exposure of persons employed in or engaged in activities involving ionizing radiation. These sources are also used to calibrate devices that can be used by first-responders in the event of a radiological emergency.

International standards, such as those developed by the International Organization for Standardization (ISO) and by the Health Physics Society (HPS), provide guidelines for personnel dosimeter and device calibrations. For example, one of the protocols involves placing such dosimeter or device on a phantom (a synthetic representation of human tissue) and exposing the dosimeter or device to determined levels of exposure (dose) and at a controlled low dose-rate (as would be encountered in actual situations). In order to attain a uniform photon emission from a ^{137}Cs source across said phantom, the source itself must be sufficiently strong to uniformly cover an area of approximately 30 cm square (the size of the reference phantom) at distances of between 200 and 300 cm. The lower half of Category 2 ^{137}Cs sources (1.0 to 50 TBq) is capable of attaining such. Category 3 (<1.0 TBq) are too weak to achieve this.

The interpretation of the device-produced isotopic yields is highly dependent on nuclear modeling. Though great improvements in the understanding of nuclear reactions have been made over the years, the *a priori* prediction of neutron capture cross sections remains very difficult. To obtain improved data for capture cross sections on unstable species, an experimental program has been launched that uses unique LANL capabilities. These include: (1) neutrons produced at Lujan Center at LANSCE; (2) a new detector system called DANCE (Detector for Advanced Neutron Capture Experiments) - a 4π 140 element BaF2 detector array to measure capture reactions; (3) capabilities for radiochemical processing of irradiation materials; and (4) a dedicated isotope separator (RSIS, Radioactive Sample Isotope Separator) of radioactive species for target preparation. To complete this integral LANL program it is necessary to have a capability to produce the isotopes required for these measurements. The LANL Isotope Production Facility can play a critical role in providing these required isotopes. This research program would, therefore, provide useful data for weapons physics as well as develop capabilities and experts in the area of isotope production and nuclear science.

4. CHALLENGES FOR THE ISOTOPE PROGRAM

The previous chapter demonstrates the broad impact, the opportunities, and some of the complexities in providing isotopes for the nation. Viewed from a high level, the challenges facing the DOE isotope program are daunting. Just considering research isotopes, the primary challenge is that the isotope program serves the research of many federal agencies. By the very nature of research, promising research opportunities change from year to year. To be effective, the program must maintain broad (and expensive) capabilities. Many of these capabilities require highly trained teams with unique expertise that cannot be easily replaced, and many have significant environment, health, and safety implications. If isotopes are used in human patients, then the Food and Drug Administration requires demonstration of and strict adherence to current Good Manufacturing Processes (cGMP). Once constructed, due to fluctuating demand, the capabilities may not be continuously in use. To be efficient, all customers and especially federal agencies must accurately project their needs, and the Department of Energy must coordinate these requests and provide feedback on actual availability. For example, the National Cancer Institute does not want to fund medical research for isotopes that will not be available, but DOE cannot plan to produce these isotopes in quantity unless they are aware in advance what isotopes and what quantities are needed. (For NIH, this issue is now being addressed in an NIH-DOE interagency working group.) The program leverages major capital investments of other parts of DOE to provide unique capabilities cost-effectively, but then is subject to changing mission priorities effecting operating schedules or even facility closure decisions outside the isotope program's control. Many radioisotopes must be used within hours or days of their production, and medical treatments require stable long-term availability. However, the isotope program currently has no accelerator facility available for the continuous production of isotopes. New capabilities can require significant capital funds and development and construction time, and if, for example, a promising new medical application fails to perform as expected in later stage trials, the demand for a particular isotope may collapse. On the other hand, if it is successful, the demand may increase by large factors, again creating a shortage in supply until successful commercialization can be achieved (Sidebar 4.1). Once a reliable commercial supply is available, DOE must leave the market. On the other hand, if a major customer pulls out of the market, the cost for all other users can increase dramatically. In one recent example, ^{252}Cf, discussed in Sidebar 4.2, the continued supply of the isotope was in jeopardy.

At the same time, foreign suppliers, in many cases subsidized by governments or capitalizing on previous government stocks, often can artificially determine the price that can be charged for an isotope. This situation greatly increases the risk for a commercial entity and inhibits U.S. commercial production, placing a greater dependency on the federal isotope program. The problem is analogous to the risk that the external controls on oil supply place on the start-up of major alternative energy technologies and where the federal government has seen the need to subsidize development. The report of the January 2009 OECD Nuclear Energy Agency workshop on the production of ^{99}Mo stated, "In addition, questions were raised regarding the long-term validity of the current economic model where the security of supply relies mainly on government-run reactors which charge only marginal costs for their irradiation services."

Sidebar 4.1: Planning for the Life Cycle of Isotope Production

Some of the challenges inherent in planning an isotope production strategy are illustrated for the high-priority alpha therapy isotopes ^{225}Ac and ^{213}Bi in the figure below. Both are currently obtained at the rate of about 500 mCi/y from milking an existing ^{229}Th source at ORNL. Projected clinical trials would require a rapidly increasing supply up to perhaps 7 Ci/y in 2012 [NO08]. Indeed, establishing the required dose would be an important element in the trials.

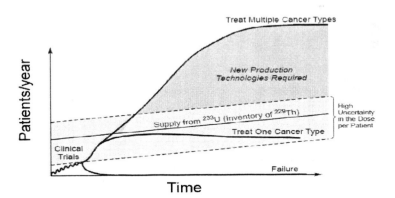

Figure. Schematic illustration of the issues of supply of ^{225}Ac/^{213}Bi depending on the success of clinical trials. C.W. Forsberg, ORNL, briefing of the U.S. DOE on March 22, 2000.

At present the most promising way to increase supply for clinical trials is to extract more ^{229}Th from stocks of ^{233}U that are mandated for waste disposal, requiring significant capital expense both for the extraction and, possibly, for the changes in the execution plan of the waste disposal activity.

If the trials fail, then the story ends. However, if trials are successful, new production technologies would be required for these alpha therapies to experience widespread use. In this case, the subcommittee was presented with future options that had the potential for such a large increase. R&D to confirm such potential is also another extremely valuable role for the isotope program.

When foreign governments subsidize research isotopes for their own researchers, the U.S. research community can be put at a significant disadvantage. Research usually begins requiring only small quantities of an isotope. If non-standard production techniques are necessary, considerable R&D may be required to supply the first batches.

The current premise of the isotope pricing policy is that the cost of research isotope production should be borne by the user, that is, by the agency funding the research. On the other hand, the DOE Office of Science operates major user facilities where the unique resources, whether they be neutrons (Spallation Neutron Source, HIFR), x-rays (i.e. Advanced Photon Source), or the world's highest energy protons (Tevatron), are available at no charge to the researcher. In those cases the selection of the research to be done is made by the DOE user facility, usually based on expert advice of an external program advisory committee (peer review), and DOE retains intellectual property rights to the work. Proprietary work at these facilities is typically charged at incremental production cost rates. There are other examples of research agencies supplying pools of unique resources to their grantees at low cost, for example, the National Cancer Institute Animal Production Program. *At present, "research" and "commercial" are defined and priced based on the isotope, not the intended use.*

Once the isotope program establishes itself as a reliable producer of a commercial isotope, care must be taken to avoid rapid fluctuations or loss of supply which could have significant economic or security impacts. This can extremely difficult in the context of normal U. S. yearly budget appropriations unless a consistent long term planning philosophy is followed.

Finally, there are very important national security issues involved in many aspects of isotope production. Highly radioactive sources can provide the material for a radiological dispersal device. Fissionable materials are frequently essential to the production of isotopes and the reactors which produce them. Non-proliferation issues must be balanced with isotope use issues. For example, the 2009 National Academy study [NR09] recommended that low-enriched uranium (LEU) should replace highly-enriched uranium (HEU) for the production of ^{99}Mo to help limit the shipment and use of the weapons-grade HEU. The isotope separation technology that is required to satisfy the need for high-purity separated isotopes could possibly be used by rogue states to create nuclear weapons; thus, much of the information on high-throughput techniques may be classified. Another aspect is the comparison of the relative benefits of extracting valuable isotopes (for example, extracting ^{229}Th from ^{233}U as a source for ^{225}Ac and ^{213}Bi) with the risks (environmental risks or the risks of diversion) of maintaining these isotopes in temporary storage facilities.

As a starting point, the "Workshop on the Nation's Needs for Isotopes: Present and Future" report highlighted several issues that are crucial to the future of the isotopes program:

- A reliable program in isotope production at DOE is crucial for the long term health of developments in medicine, basic physical and biological sciences, national security and industry.
- Many of the isotopes in domestic use are produced only by foreign suppliers, often a single or limited number of suppliers. This makes the isotope supply vulnerable to interruption or large price fluctuations beyond the control of the United States.
- Affordability is an important issue for research isotopes.

- The production capability of the NIPA program relies on facilities that are operated by DOE for other primary missions.
- There is a pressing need for more training and education programs in nuclear science and radiochemistry to provide the highly skilled work force for isotope application.
- The DOE isotope program with the resources that it has available to it today cannot fulfill the broad challenges and needs for current and future demands of the nation for isotopes.

Sidebar 4.2: ^{252}Cf, a Valuable Neutron Source

The radioisotope ^{252}Cf is an intense neutron emitter that can be packaged in compact, source capsules. Although this radioisotope decays mostly by alpha emission, the ~3% spontaneous fission branch results in a neutron emission of $1.14 \times 10^6 \text{tg}^{-1}$. The half-life of ^{252}Cf is 2.645 years, corresponding to a specific activity of 0.536 mCi/µg. The neutron energy spectrum of ^{252}Cf, with most probable energy of 0.7 MeV and an average energy of 2.1 MeV, is similar to that of a fission reactor. Therefore, small portable ^{252}Cf neutron sources can provide an ideal nonreactor source of neutrons for lower-flux applications. Larger masses of ^{252}Cf (>0.1 g) can approach reactor capabilities for many applications.

There are many uses of ^{252}Cf in research and national security, some of which are described in Chapter 3.B and 3.C. In addition, there are many commercial uses for this reliable, cost- effective neutron source. A major industrial use of this radioisotope is in prompt gamma neutron activation analysis (PGNAA). This method is used in the analysis of coal, cement, and minerals, as well as for detection of explosives. Sources of ^{252}Cf are also used in neutron radiography, nuclear waste assays, reactor start-up sources, calibration standards, and cancer therapy.

There are only two locations in the world capable of producing ^{252}Cf, the Research Institute of Atomic Reactors (RIAR) in Dimitrovgrad, Russia, and the Oak Ridge National Laboratory (ORNL), USA. The ^{252}Cf produced at ORNL supplies ~75% of the world's market.

In the spring of 2007 with the withdrawal of a major ^{252}Cf customer, NNSA, from the market, ongoing sales could not support the significant up-front costs to prepare the production targets during the 1-2 year time-scale required from initiation until the final sources are available for sales. The future of ^{252}Cf production in the U.S. was in question. The Department of Energy (DOE) sought input from both ORNL and key industrial partners to develop a solution to maintain the production of ^{252}Cf. The industry projected their ^{252}Cf demands for the next five to ten years and ORNL prepared an analysis of the baseline costs and equipment upgrades necessary to meet the projected industry's demand. With this information in hand, during the last week of May 2009, the DOE was able to enter into a unique contract with the industry partners that will ensure the continuous supply of ^{252}Cf, with industrial subscribers providing a significant fraction of the up-front costs. The first ^{252}Cf processing campaign under this new contract is currently underway with ^{252}Cf expected to be available for source production by July 2009. As a result of the new arrangement, the supply is assured, though a significant increase in the price of the ^{252}Cf is anticipated. The final price has not been made public by the commercial distributers.

The workshop by design did not address the relative priorities for uses for various isotopes. Setting priorities between various disciplines and end users is clearly another major issue. With this input, the Subcommittee organized its discussions around the following issues:

- How should DOE maximize the effectiveness of the limited resources of the program to fulfill its mission (as expressed in the FY09 President's budget request) to support the research and development and production of isotopes and to make them more readily available to respond to the needs of the nation?
- What new capabilities need to be added to the program to support the known needs?
- How can new and improved techniques be developed to increase the variety and quantities of isotopes produced and make isotope production more cost effective?
- How can the required cadre of highly skilled personnel be cultivated to ensure a future supply of isotopes?

This latter issue involves both ensuring stable support for the existing highly specialized individuals and training and work experience for the next generation to produce and develop isotopes.

In the next seven chapters, these issues will be developed and recommendations made. Chapters 5 to 8 address the four major production techniques. Chapters 9 and 10 consider the intellectual capital and skilled workforce and Chapter 11 addresses program operations. The implications on the budget for the program are presented in Chapter 12, and then Chapter 13 concludes by collecting the recommendations of the previous sections.

Sidebar 4.3: Molybdenum-99: A Major Concern

^{99}Mo is used to produce ^{99}Mo/^{99m}Tc generators (a generator technology developed at Brookhaven National Laboratory) for use in nuclear medicine. ^{99m}Tc is the most widely used radionuclide in nuclear medicine, both for detection of disease and for the study of organ structure and function. More than 15 million procedures are performed each year in the U.S. using ^{99m}Tc. There are more than 40 diagnostic tests available to physicians using ^{99m}Tc.

While ^{99}Mo was originally produced from direct neutron capture reactions in reactors such as MURR, Cintichem, Inc. began production using neutron-induced fission reactions with ^{235}U targets of uranium highly enriched (>20%) with (HEU) in 1980 in Tuxedo, New York. This reactor was shut down in 1989 following concerns involving tritium contamination and was later decommissioned. The private sector was not willing to assume the financial and regulatory risks associated with building and operating a new reactor facility. Cintichem did arrange a long-term supply agreement with the Canadian company Nordion (later MDS Nordion), to supply ^{99}Mo to U.S. technetium generators (Amersham [now GE Healthcare], Mallinckrodt and Dupont).

DOE purchased the rights to Cintichem's production technology and agreed to take back their associated waste. DOE investigated using various existing reactors and hot-cells at LANL and SNL to bridge the gap in domestic supply. In 1999, DOE completed conversion of facilities at Sandia for medical isotopes but did not start ^{99}Mo production. With the entry of European suppliers into the U.S. market in 1998 and the commitment by Atomic Energy of Canada Limited to build two new ^{99}Mo production reactors, the supply of ^{99}Mo appeared more diverse and reliable. Moreover, the private sector believed it would not be economically competitive to enter the market using the SNL reactor operations.

Currently, there are five major producers of ^{99}Mo internationally, and all currently use highly enriched uranium targets to produce ^{99}Mo. The U.S. is the primary supplier of HEU for the world's ^{99}Mo production, including about 15 kg/year to Canada. Canada built the two new Maple reactors to replace their aging NRU reactor with a total capacity roughly equal to the current worldwide demand. However, these reactors did not perform as designed. The cause for the discrepancies was not determined, and Atomic Energy Canada Limited has halted work on the project. In 2006, 2007 and 2009, the NRU reactor had unplanned shutdowns. Coupled with shutdowns at other international reactors, these problems led to serious shortages in the ^{99}Mo supply, affecting physicians and patients.

There has been a concerted effort over the last two decades to eliminate the use of HEU because of serious concerns over non-proliferation. HEU used in medical radionuclide production is typically enriched at levels between 20-90% ^{235}U, and the production reactors typically use HEU fuel. The DOE HEU elimination efforts are carried out under the DOE/National Nuclear Security Agency Global Threat Reduction Initiative. A 2009 report of the National Academy of Sciences [NR09] concluded that it was feasible to use LEU targets for ^{99}Mo production with an acceptable increase in cost.

The ^{99}Mo producers are supportive of the conversion from HEU to low-enriched uranium (LEU). However, the ^{99}Mo producers need time to convert from HEU to LEU.

New radiochemical procedures using LEU targets and waste stream protocols need to be developed before the current ^{99}Mo production can be replaced by LEU. The use of LEU has been demonstrated by several small scale operations around the world. No large scale production of ^{99}Mo by this technique has been demonstrated as of yet.

The DOE/NNSA has the lead for the Department on the ^{99}Mo issue. Congress has mandated a report from DOE within one year containing the findings of the NAS study and disclosing the existence of any commitments from commercial suppliers to provide the domestic requirements for medical isotopes without HEU by 2013.

As a result of the intense ongoing activity and the active investigations of specific commercial alternatives, the NSACI subcommittee will not enter this debate at this time or make any specific recommendations in this area. The subcommittee does consider the issue to be a critical one and does go on record with its concern that it must be addressed expeditiously.

Major Concern

The supply of ^{99}Mo, the isotope used to generate the radioactive isotope most frequently used in medical procedures, is of great concern. Recent disruptions in international supply demonstrate the vulnerability of the nation's health care system in this area. The nation must address this vulnerability. At the present time, the isotopes program does not produce ^{99}Mo. With the non-proliferation issues associated with the transport and use of the highly-enriched uranium currently used for ^{99}Mo production, DOE/NNSA has the lead responsibility in this area and is actively investigating options for ^{99}Mo commercial production. The subcommittee chose to refrain at this time from inserting itself into the intense activity underway, but reiterates the importance of the issue.

5. Stable Isotopes

Stable isotope applications are widespread in the fields of basic and biomedical research and in medical diagnosis and treatment. Sidebar 5.1 presents the history of isotope labeling in living systems. Further, isotope supplies are essential for diverse applications in industry and national security (Figure 5.1). Three large general and two more specialized application areas emerge.

1) The "light" stable isotopes of hydrogen, carbon, nitrogen, oxygen, and sulfur are used for the study of virtually all aspects of the chemistry, basic biochemistry and clinical biochemistry, and metabolism of organic molecules. These applications run

into many thousands and are far too numerous to detail. Large amounts of ^{2}H2O are also required for heavy-water reactors and as a moderator for neutron sources such as the Spallation Neutron Source at ORNL. Helium-3 is used medically for hyperpolarized gas *in vivo* magnetic resonance imaging studies of pulmonary ventilation, but is particularly important for its non-medical uses in fusion studies, cryogenic applications, and neutron detection.

2) The stable isotopic nuclides of heavier elements are used for innumerable agricultural, nutritional, industrial, environmental, ecological, and materials science applications. Nitrogen cycles in worldwide agricultural research can only be studied with ^{15}N. Essential nutrients in man include F, Na, Mg, P, S, Cl, K, Ca, Cr, Mn, Fe, Co, Ni, Cu, Zn, Se, Mo, and I. Some of these elements (e.g. F, Na, P, Mn, I) are monoisotopic and, thus, not amenable for use in tracer studies. The remainder, however, have stable nuclides that are critically necessary for investigation of the requirements and metabolism of these indispensible nutrients in humans and animals [TU06, ST08, FA02]. Similarly, stable isotopes are used to answer immensely diverse environmental and ecological questions at every level on a global scale (e.g. [CH09, LY07, RA09, WA91]), as well as probe subjects of archeological and popular historical interest [LY07]. Geological and paleontological applications include the use of Pb, Sm, Rh, Os, Pa, Th, and Sr, in addition to the lighter stable isotopic nuclides C, N, O, and P.

3) An additional significant use of stable isotopes is as targets for the production of critical radioisotopes (Table 5.1). This use is, in fact, the principal world demand for stable isotopes with thallium-203 sales leading those of all others. ^{15}N is the source of ^{15}O, and ^{18}O is the target for production of ^{18}F, all positron emitting radionuclides used in PET scanning. Additional medical uses include the production of ^{123}I from ^{123}Te for thyroid scanning and ^{103}Pd from ^{104}Pd for brachytherapy seeds. Various additional stable isotopes serve as targets for production of a large variety of radioisotopes for industrial applications and calibration sources. In nuclear physics research, targets or intense beams of stable isotopes with low natural abundances are often essential, e.g. for the production of super-heavy elements or to produce beams of rare radioactive isotopes far from stability. For example, each year nuclear physics laboratories are a significant consumer of separated ^{48}Ca (which has a natural abundance of only 0.2%).

Sidebar 5.1: Stable Isotope Tracers in Living Systems

Seventy-five years ago, Rudolf Schoenheimer proposed that biochemical substances labeled with deuterium could be used to trace metabolic events *in vivo* (KE01, SC35A, SC35B]. In an elegant and extensive series of papers published over the next six years, he and his associates then proceeded to define the dynamic state of body constituents using deuterium labeled substrates [SC40, SC42, GU 91, SI02]. Presciently, he noted that "the number of possible applications of this method appears to be unlimited" [SC35A] and that "the use of a carbon isotope ^{13}C instead of deuterium would open some fields which cannot be attacked with deuterium...." while "nitrogen isotopes could undoubtedly open a wide field of investigation of the nitrogen metabolism." [SC35B].

During the same period, before ^{14}C and ^{3}H were used as biological tracers, Cohn and Greenberg traced the metabolism of phosphorus in rats with ^{32}P [CO38] and Hahn, *et al.* studied the metabolism of iron in dogs with radioactive iron [HA39] produced by Lawrence's cyclotron in Berkeley. By the late 1940's, the widespread availability of the radiotracers ^{14}C and ^{3}H for biological studies and the relative ease of their use, led to the virtually uniform adoption of radiotracers for the study of biological processes *in vivo*. An exception was the very limited use of ^{15}N for the study of nitrogen metabolism, since there is no long-lived radiotracer for nitrogen. Although stable mineral isotopes were being produced by the Calutrons at Y-12/Oak Ridge National Laboratory, the lack of readily available analytical methods severely curtained the use of stable isotopic mineral tracers for nearly two decades [TU06]. Likewise, the lack of production of enriched ^{13}C, ^{15}N, and ^{18}O severely curtailed the use of these stable isotopes for research in any biological systems. Deuterium was available, but its relative expense and difficulty of analysis limited its routine application.

Thus, until about 1970, virtually all tracer research in living systems was conducted with radiotracers, despite the mounting ethical concerns about using these tracers for studies in humans. Thereafter, however, this situation changed dramatically. Initial large scale production of ^{13}C, ^{15}N, and ^{18}O by the ICONS Program at Los Alamos National Laboratory, subsequent production by commercial sources, and advances in mass spectrometric and magnetic resonance approaches for measuring these isotopes, allowed stable isotope tracers to supplant their respective radiotracers for biological use. Similarly, developments in thermal ionization mass spectrometry and inductively coupled plasma mass spectrometry [TU06, ST11] permitted the general application of stable isotopic mineral tracers, produced by the Calutrons, to the solution of biological problems. The success of these combined efforts has been so dramatic that today virtually all tracer studies in humans are performed with stable isotopes, rather than with radiotracers.

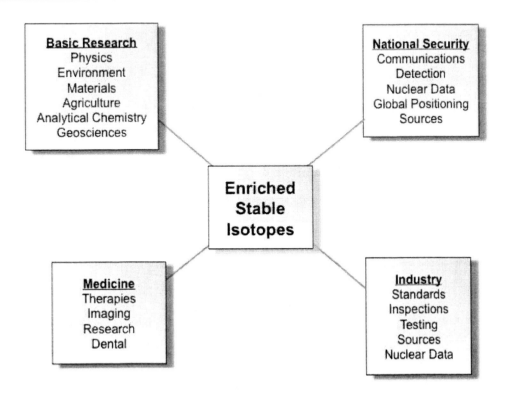

Figure 5.1. Uses of stable enriched isotopes.

Table 5.1. Selected enriched stable isotopes and derived radioisotopes [IM95]

Stable Isotope Target	Radioactive Product
Cadmium-112	Indium-111
Carbon-13	Nitrogen-13
Chromium-50	Chromium-51
Germanium-76	Arsenic-77
Lutetium-176	Lutetium-177
Nickel-58	Cobalt-57
Nitrogen-15	Oxygen-15
Oxygen-18	Fluorine-18
Palladium-102	Palladium-103
Platinum-198	Gold-199
Rhenium-185	Rhenium-186
Samarium-152	Samarium-153
Strontium-88	Strontium-89
Thallium-203	Thallium-201
Xenon-124	Iodine-123
Zinc-68	Gallium-67, Copper-67

A specialized but particularly large quantity application arises in the search for physics beyond the standard model in neutrino-less double beta decay. These experiments which have high priority in the nuclear and elementary particle communities [NS07] may require ton quantities of selected enriched isotopes such as ^{76}Ge.

The lithium isotopes, ^{6}Li and ^{7}Li, represent another set of specialized but large quantity (reaching tens of thousands of kilograms compared to current sales of ~20 kg/yr) applications. ^{6}Li is important for nuclear weapons purposes and for future fusion reactors. High-purity ^{7}Li has potential uses as a part of reactor shielding or for the cooling systems for certain molten salt reactors, where ^{6}Li contamination would lead to unacceptably high levels of tritium production. The existing technology used for lithium separations involves mercury amalgams, with significant environmental and health issues.

Production, Supply, and Availability

Stable isotope production was once a major endeavor for the federal government as an outgrowth of uranium enrichment efforts. Thirty of the "Calutron" electromagnetic separators at Y-12/ORNL started production of stable isotopes (See Figure 5.2) in 1945, extending eventually to over 250 stable isotopes, and pioneered the way for many new innovative uses. In 1998, production operations ceased, in large part due to the requirement for full cost recovery and the price undercutting through sales of subsidized foreign stockpiles. The Calutron facilities have now been shut down for many years and the effort, environmental issues, and major capital investment that would be required to resume production with these five decade old devices in today's regulatory environment is viewed as prohibitive. It is difficult to estimate the value of the existing pool of isotopes which represents a valued asset to the nation. If all were sold at list price, it would be very high, but this does not represent a true market value. It is being consumed by current sales of about $0.5-1M per year. Based on actual sales, an independent auditor places the commercial value at $3.5M. Over the last five to six years, ORNL has sold stable isotopes to approximately 100 unique users per annum, on average. There are no reliable, documented data on the supply from foreign sources except the aggregate numbers in the U.S. International Trade Commission ITS-01 report [ITS09]. At ORNL, the bulk of nuclides requested are for amounts in the hundreds of milligrams per year range. Approximately 15 nuclides are supplied annually in the range of 1 to 10 grams, approximately 5 nuclides are sold at annual amounts in the 10 to 20 gram range, and about the same numbers of nuclides are requested annually in more than 20 gram amounts.

Although significant stock of the stable isotopic nuclides of the heavier elements exists in current DOE inventory, some nuclide stocks have been exhausted, and there is less than a 20 year supply of many others (Table 5.2). The great bulk of the stable mineral isotopes used for human research are supplied by Russia, and there is great concern for future availability (which extends back many years [AB92]). It is not an exaggeration to say that research and clinical studies of essential mineral nutrient metabolism in man, as well as the broad array of environmental and ecological studies, will come to a complete halt if the supply of these elements is curtailed.

Active production of light stable isotopes in the United States is currently primarily performed by the private sector with the exception of Helium-3 which is produced at the Savannah River facility operated by the NNSA as a by-product of tritium decay (See Chapter 8). For the purposes of this report *production* refers to a process whereby a stable isotope of an element is both separated and enriched to a useable level which is typically above 90 atom%.

The widespread use of ^2H, ^{13}C, and ^{18}O throughout basic and clinical biochemical research has made commercial production of these isotopes feasible, and industry sources are readily available. ^{15}N demand is also met currently by industry sources, but it is not available domestically, and there is no domestic generator of new inventory. The latter is, potentially, no trivial problem because nitrogen is an indispensible dietary nutrient, especially in its role as the essential nutrient in amino acids, the building blocks of proteins. Thus, since there is no long- lived radiotracer alternative, an absence of ^{15}N would curtail essentially all human studies of nitrogen metabolism. Similarly, ^{18}O is absolutely essential for the production of the widely used positron emitter ^{18}F and is a critical nuclide for use in the "doubly labeled water" method, the only method available for measuring energy expenditure of animals and humans in the free living state. It is not merely academic to be concerned about future interruptions in scientific activities due to shortages of commercial isotope supplies. Following termination of the Los Alamos source of ^{18}O in 1989-1990, commercial suppliers could not keep up with demand for approximately one year. More recently, from approximately 2003-2005, heavy demand of 18O for PET scanning applications caused the widespread unavailability of this nuclide for use by the nutrition community for energy expenditure measurements. The latter shortage was anticipated by the North American Society for the Study of Obesity (now The Obesity Society) when it issued a report on the "Supply and Demand of Oxygen-18 Water" in 1999 [AH99]. Currently, there are no publically available hard data on research demand or commercial production capacity for ^{18}O, but a significant fraction comes from non-U.S. producers. The nutrition community consensus is that a significant shortfall would exist if foreign sources of ^{18}O were made unavailable. Similarly, a significant amount of deuterium used in the U.S. comes from foreign sources, and the isotopes of nitrogen, halogens, and noble gasses are available only from foreign suppliers.

The essential attributes for a scientific supply of commercial stable isotopes, like those for radioisotopes, include predictability, reliability, and long-term sustainability of product supply, and competitive pricing for the research community. Dependence on foreign sources for essential stable isotopes violates these principles and potentially puts the Nation's future research, diagnostic, therapeutic, and industrial activities at serious risk. This risk has been recognized and highlighted by every expert committee constituted to study the problem over the last two decades, including a detailed recent assessment that included recommendations for the future strategies necessary to sustain the U.S. isotope supply [RI05].

There are eight primary techniques for stable isotope separation.

- Electromagnetic separation
- Gaseous diffusion
- Gas centrifuges
- Thermal diffusion
- Distillation
- Chemical exchange
- Plasma separation
- Laser separation

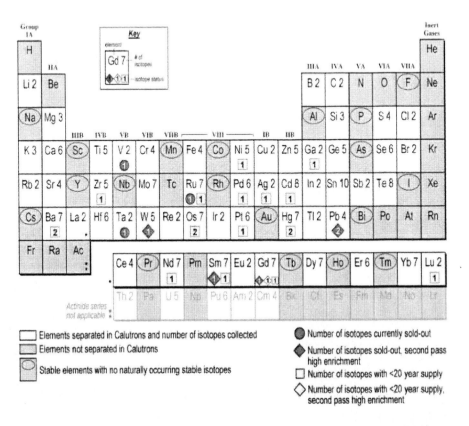

Figure 5.2. Elements with isotopes enriched by Y-12/ORNL Calutrons and current availability.

Isotopes for the Nation's Future: A Long Range Plan 83

The methods of isotope separation and enrichment employed by the private sector companies are distillation, chemical exchange, and thermal diffusion. The private companies which have these capabilities are Cambridge Isotope Laboratories, Eagle Pitcher, Isotec (Sigma Aldrich), and Spectra Gases. They produce the isotopes of carbon-13, oxygen-17, 18, and boron-10, 11, with capacity in the metric ton range, and they offer a wide variety of compounds labeled with these isotopes. Supply is not an issue for any of these particular isotopes. Additionally, Isotec has a set of thermal diffusion columns which can be used for production of gaseous isotopes. These systems are not competitive in cost of production to those using the cryogenic centrifuges employed by foreign manufactures but do provide some domestic capability for some isotopes such as krypton, and xenon. Additionally, Spectra Gases has systems capable of enriching nitrogen-15, but the demand for this isotope is currently met by foreign entities at extremely reasonable prices.

Table 5.2. Remaining inventory of selected stable isotopic nuclides in short supply at Oak Ridge National Laboratory

Isotope	Years Remaining Inventory
Gadolinium-154, Second Pass	2.5
Gallium-69	3.7
Nickel-62	3.9
Osmium-187	5.2
Lutetium-176	5.5
Ruthenium-99	6.3
Osmium-186	7.5
Barium-136	7.6
Neodymium-150	7.9
Mercury-204	10.2
Cadmium-106	10.7
Mercury-202	11.5
Palladium-106	12.6
Silver-109	14.3
Zirconium-94	18.5
Barium-137	19.0
Samarium-149	19.6
Gadolinium-157	0.2
Platinum-195	12.0
Gadolinium-157, Second Pass	0.0
Lead-204, Second Pass	0.0
Lead-207, Second Pass	0.0
Ruthenium-96	0.0
Samarium-150, Second Pass	0.0
Tantalum-181	0.0
Vanadium-51	0.0
Tungsten-180, Second Pass	0.0

Plasma isotope separation is a tool that has been used successfully in the U.S. weapons programs and is capable of producing specific isotopes at medium enrichments. These systems have been dismantled and are no longer available; however, the technology is useful to an enrichment program if integrated with electromagnetic separators. A private company, Nonlinear Ion Dynamics (NID), LLC, in California, has demonstrated a plasma isotope enrichment system and has operated under an SBIR grant and private funding.

The vast majority of the 250 naturally occurring stable isotopes are primarily made up of alkalis, alkali-earth, and metal stable isotopes and require the use of electromagnetic separators which are no longer in use within the U.S. The 220 stable non-gaseous isotopes are not currently produced in the U.S. The reasons for this are several but include the following:

1) Most are only used in research applications in very limited quantities which will not attract private sector investment to manufacture them.

2) Most require separation and enrichment by means of either electromagnetic or gas centrifuge separators. These systems are not operational in the U.S., are expensive to

3) Many of the isotopes produced by electromagnetic separators are currently inventoried at ORNL in sufficient quantities to support limited research; however, a number of isotopes (See Table 5.2) are no longer in inventory and/or are well below levels to sustain research even in the short term.

4) Foreign supply of the stable isotopes requiring centrifuges is currently meeting demand in most cases. However, in a few instances the foreign supply is not meeting demand, e.g. ^{136}Xe, and ^{76}Ge.

5) The thermal diffusion method used for the separation and enrichment of the rare gas isotopes of argon, neon, krypton, and xenon is an expensive method of production.

6) Plasma isotope separators appear very promising for high-throughput applications but require more research and development to be used as a production tool.

7) Laser isotope separators appear prohibitively expensive to operate and currently limited in their scope of production for stable isotopes.

The alkalis, alkali-earth, and metal stable isotopes are essential to current research in health care and nutrition studies, which manipulate biochemistry at the cellular and sub-cellular level to prevent disease, as well as to offer personalized detection and treatment. The research cannot be done without the assurance of an ongoing supply of these isotopes. Some of these (Table 5.2) are already in short supply or no longer available and the only demonstrated method of production and separation for these stable isotopes is electromagnetic separators.

Research in the U.S. that uses stable and enriched isotopes is strategically important and in the Nation's interest. Therefore, the domestic capability for the production of these stable isotopes is strategically important to the U.S. in order to assure the continuation of research activities.

The subcommittee concludes that it is essential that the U.S. reestablish the base production capability for stable isotope separation. Not surprisingly, this recommendation reiterates those of many previous reports [IM95, NE00, RI05]. Currently, electromagnetic separation seems to be the only general applicable technique capable of the very high enrichments needed in some applications. Based on the high-priority research needs identified in the Subcommittee's first report (See Tables 8 and 9 of [NS09] reproduced in Appendix 7), some separation capability for radioactive isotopes would also be valuable. While the subcommittee did not attempt to define such a facility in detail and did not receive any specific proposals that it would endorse, the general parameters of such a capability should be multiple separators, each of a capacity similar to one Calutron (~100 mA ion current of feedstock). A typical configuration might be four separators, two in production for stable isotopes, one separator dedicated to the special problems of radioactive material, and one in the process of setup, cleanup, maintenance, or R&D. Such a configuration is scaled to annual replacement of recent average yearly isotope sales. It would allow some economies of scale and shared expertise but also provide flexibility depending on demand.

While electromagnetic separators remain the best choice of a general purpose isotope separation technology, plasma techniques show considerable promise, particularly for high volume applications such as neutrino-less double beta decay. Continued R&D is certainly warranted, and in the future this technology may be an integral part of the stable isotope production capability portfolio.

An important issue for stable isotope production is the potential use of these technologies for weapons of mass destruction. As such, much of the forefront work in this area may be classified, and the technology is subject to security and export controls. Such considerations were beyond the scope of this subcommittee. Clearly, it would be advantageous to be able to take advantage of such classified information, and experts knowledgeable in these areas need to be involved in the review of any future proposed capabilities. Security issues may also limit the possible choices of sites for new separator facilities to sites with appropriate security measures.

Finally, the distribution of a broad variety of isotopes requires the on-site presence of chemical and materials processing laboratories such as are currently available at ORNL. In order to make the stable isotope supply useful to customers, the following services are needed.

- Metallurgical, ceramic, and high vacuum processing methods
- Pyrochemical conversion: oxide to high-purity metal
- Arcmelting and alloying hot and cold rolling
- Preparation of cold-rolled foils from air-reactive metals
- Drop casting
- Wire rolling/swaging (hot or cold)
- Target fabrication

Recommendations

Program Operation
Maintain a continuous dialogue with all interested federal agencies and commercial isotope customers to forecast and match realistic isotope demand and achievable production capabilities.

Support a sustained research program in the base budget to enhance the capabilities of the isotope program in the production and supply of isotopes generated from isotope separators.

Major Investments in Production Capability

Construct and operate an electromagnetic isotope separator facility for stable and long-lived radioactive isotopes.

It is recommended that such a facility include several separators for a raw feedstock throughput of about 300-600 milliAmpere (10-20 mg/hr multiplied by the atomic weight and isotopic abundance of the isotope). This capacity will allow yearly sales stocks to be replaced and provide some capability for additional production of high-priority isotopes.

6. ACCELERATOR BASED ISOTOPE CAPABILITIES

Isotopes produced at accelerators are typically neutron deficient and are made with cyclotrons or linear accelerators by high-current proton, deuteron, or alpha particle bombardment. Accelerator isotope applications generally complement reactor isotope applications, and accelerator isotopes usually decay by ʙ, ˣ positron emission or electron capture. The accelerator beam parameters, especially beam energy and beam current, are important considerations in the production of isotopes. Beam energy determines what isotopes are produced (and by what nuclear reaction), and beam current determines how much is produced. Low-energy cyclotrons (<30 MeV) are generally used to produce short-lived isotopes (^{11}C, ^{15}N, and ^{18}F) that are used in clinical positron emission tomography (PET) and PET R&D. However, many other isotopes can be made at lower energies (See Table 6.1). Several commercial isotopes are produced in 30 MeV cyclotrons operated by industrial isotope producers and radiopharmaceutical manufacturers, e.g. ^{111}In, ^{201}Tl, ^{67}Ga and ^{123}I . Higher-energy accelerators are usually operated by government laboratories and make products that require the higher energy, e.g. ^{82}Sr. The discussion below focuses on all of these capabilities, including low-energy university cyclotrons, commercial cyclotrons, the Department of Energy (DOE) production capabilities on higher-energy accelerators at Brookhaven National Laboratory and Los Alamos National Laboratory, and international accelerator collaborators.

Scope of the Accelerator Production Capability

Commercial Cyclotrons (Usually 30 MeV)

Radiopharmaceutical manufacturers and radioisotope producers have operated 30 MeV cyclotrons to insure the availability of commercially viable radioisotopes to support radiopharmaceutical applications and clinical nuclear medicine. Most of these radiopharmaceutical manufacturers, including Covidien, GE Healthcare, Lantheus, MDS Nordion, and NuView (whose capabilities include a linac with 70 MeV potential) produce radioisotopes to support their radiopharmaceutical businesses. These commercial radioisotopes include ^{201}Tl, ^{111}In, ^{123}I , as well as other isotope products. The radio-pharmaceutical manufacturers do this to insure the reliable availability of radio-pharmaceuticals for the clinical practice of nuclear medicine. Most of these commercial producers have multiple cyclotrons for this production, but since the cyclotrons are in various

Isotopes for the Nation's Future: A Long Range Plan

stages of life cycle, future capital investments for each manufacturer will be different. There are more than a dozen of these commercial cyclotrons operating in the U.S.

Table 6.1. Isotopes that can be made with lower energy reactions

Isotope	Half-Life	Production Reactions	Beam Energy (MeV)
Be-7	53.3d	$^6Li(d,n)$, $^7Li(p,n)$	10 (d), 10 (p)
C-11	20.4m	$^{11}B(p,n)$, $^{10}B(d,n)$	20 (p), 10 (d)
N-13	9.96m	$^{12}C(d,n)$	10
O-15	2.04m	$^{14}N(d,n)$	20
F-18	1.83h	$^{18}O(p,n)$, $^{20}Ne(d,a)$, $^{18}O(a,d)$	20 (p), 10 (d)
Na-22	2.62y	$^{22}Ne(p,n)$, $^{20}Ne(a,d)$, $^{24}Mg(d, a)$	20 (p), 20 (d)
Al-26	7.2E05y	$^{26}Mg(p,n)$	20
P-32	14.3d	$^{32}S (d,2p)$	?
Sc-46	83.8d	$^{45}Sc(d,p)$, $^{46}Ti(d,2p)$	?
Sc-47	3.34d	$^{48}Ca(p,2n)$	30
Ti-44	47y	$^{45}Sc(p,2n)$	30
V-48	16.0d	$^{48}Ti(p,n)$	20
V-49	330d	$^{48}Ti(d,n)$	15
Cr-51	27.7d	$^{51}V(p,n)$	20
Mn-52	5.59d	$^{52}Cr(p,n)$	20
Fe-55	2.73y	$^{55}Mn(p,n)$	20
Co-55	17.5h	$^{56}Fe(p,2n)$	30
Co-56	77.7d	$^{56}Fe(p,n)$	20
Co-57	272d	$^{56}Fe(d,n)$, $^{58}Ni(p,2n)$ ^{57}Cu (decay)	10 (d), 30 (p)
Co-60	5.27y	$^{59}Co(d,p)$	?
Cu-61	3.41h	$^{60}Ni(d,n)$	15
Cu-64	12.7h	$^{64}Ni(p,n)$	20
Cu-67	2.58d	$^{70}Zn (p,a)$, $^{67}Zn(d,2p)$	25 (p), ?
Zn-62 /	9.26h/9.74	$^{63}Cu(p,2n)$	30
Zn-65	244d	$^{65}Cu(p,n)$	20
Ga-67	3.26d	$^{66}Zn(d,n)$	15
Ge-68	271d	$^{69}Ga(p,2n)$	30
As-72	26h	$^{72}Ge(p,n)$	35
As-76	1.10d	$^{75}As(d,p)$	30
As-77	1 .62d	$^{76}Ge(d,p)$ ^{77}Ge (decay)	?
Se-75	120d	$^{75}As(p,n)$	20
Br-75	1.63h	$^{74}Se(d,n)$	15
Br-76	7.2h	$^{76}Se(p,n)$	20+
Br-77	2.38d	$^{78}Se(p,2n)$	30
Br-82	1.47d	$^{81}Br(d,p)$	30
Rb-81	4.58h	$^{82}Kr(p,2n)$	30

Table 6.1. (Continued)

Isotope	Half-Life	Production Reactions	Beam Energy (MeV)
Rb-83	86.2d	$^{84}Kr(p,2n)$	30
Rb-86	18.7d	$^{85}Rb(d,p)$	30
Sr-85	64d	$^{85}Rb(p,n)$	20
Sr-89	50.6d	$^{88}Sr(d,p)$	30
Y-88	107d	$^{88}Sr(p,n)$, $^{88}Sr(d,2n)$	20 (p), 30 (d)
Y-90	2.67d	$^{89}Y(d,p)$	30
Zr-88	83.4d	$^{89}Y(p,2n)$	40
Zr-89	3.27d	$^{89}Y(p,n)$	20
Tc-95, Tc-	20.0h, 61d	$^{95}Mo(p,n)$	20
Tc-96	4.3d	$^{96}Mo(p,n)$	20
Mo-99 /	2.75d /	$^{96}Mo(a,n)$	?
Rh-105	1.47d	$^{104}Ru(d,p)$, $^{104}Ru(d,n)$(decay)	20, 15
Pd-103	17.0d	$^{103}Rh(p,n)$	20
Pd-109	13.7h	$^{108}Pd(d,p)$	20
Cd-109	462d	$^{109}Ag(p,n)$, $^{109}Ag(d,2n)$	20 (p), 30 (d)
In-111	2.81d	$^{110}Cd(d,n)$ or $^{111}Cd(p,n)$	15 (d), 20 (p)
I-123	13.2h	$^{122}Te(d,n)$, $^{123}Te(p,n)$, $^{124}Xe(p,2n)$ (decay)	15 (d), 20 (p), 35 (p)
I-124	4.18d	$^{124}Te(p,n)$	20
I-125	60.1d	$^{124}Te(d,n)$, $^{124}Xe(d,p)$(decay)	15, 25
Xe-127	36.4d	$^{127}I(p,n)$	20
Ce-139	138d	$^{139}La(p,n)$	20
Pr-142	19.1h	$^{141}Pr(d,p)$	20
Gd-153	242d	$^{153}Eu(p,n)$, $^{153}Eu(d,2n)$	20
Gd-159	18.6h	$^{158}Gd(d,p)$	20
Dy-165	2.33h	$^{164}Dy(d,p)$	20
Ho-166	1.12d	$^{165}Ho(d,p)$	20
Re-186	3.78d	$^{186}W(p,n)$, $^{186}W(d,2n)$	20 (p), 25 (d)
Os-191	15.4d	$^{190}Os(d,p)$	20
Ir-191m	4.94s	Daughter Os-191	
Ir-192	73.8d	$^{191}Ir(d,p)$, $^{192}Os(p,n)$, $^{192}Os(d,2n)$	25 (d), 20 (p), 25 (d)
Ir-194	19.2h	$^{193}Ir(d,p)$	25
Au-198	2.69d	$^{197}Au(d,p)$	25
Au-199	3.14d	$^{198}Pt(d,n)$	20
Hg-197	2.67d	$^{197}Au(p,n)$	20
Pb-203	2.17d	$^{203}Tl(p,n)$	20
Bi-206	6.24d	$^{206}Pb(p,n)$	20
Bi-207	32y	$^{207}Pb(p,n)$	20
At-211	16.2h	$^{209}Bi(a,2n)$	26

Isotopes for the Nation's Future: A Long Range Plan

In addition to the production of these commercial radioisotopes, 30 MeV cyclotrons could be very useful to meet research isotope availability missions because more than 90 % of research isotopes can be produced at energies below 30 MeV [QA82, QA01, QA04, RU89]. Table 6.1 lists the isotopes that can be produced at or below 40 MeV and the nuclear reactions that are used to produce the isotopes ("?" indicates the precise optimum beam energy needs to be determined). Because the commercial cyclotrons are usually fully subscribed, they are usually not available for research isotope production.

University PET Cyclotrons

For many years universities have operated low-energy cyclotrons (typically <20 MeV) for the production of short-lived PET radioisotopes, including ^{18}F, ^{11}C, and ^{15}O for research and development. Since the evolution of PET from a research tool to a clinical tool for nuclear medicine diagnosis, especially oncology diagnosis and treatment efficacy, there has been an explosion of these low-energy cyclotrons at hospitals, nuclear pharmacies, and research centers in the U.S. and internationally. While these cyclotrons are well utilized for the production of these short-lived PET radioisotopes, there is under-utilized capacity at most of these facilities (for example, night shifts when patients are not being treated) that could be used to enhance future research isotope availability. This under-utilized capacity is an opportunity for the isotope program. However, numerous logistical issues, from chemical processing, to regulatory compliance, to transportation logistics, to distribution, will need to be solved. Washington University in St. Louis has had an NIH/NCI grant for several years to use their low-energy cyclotron to make research radioisotopes, including ^{64}Cu and ^{86}Y, for the nuclear medicine research community (Sidebar 9.1). Several low-energy cyclotrons have also been installed in nuclear pharmacies and hospitals. Many hospitals and universities have partnered with commercial nuclear pharmacies to market products produced at these facilities. As discussed later in the chapter, better utilization of these university cyclotrons could significantly increase the availability of the research isotopes that are best produced at lower energy.

Alpha particle beams of about 26 MeV in energy are required to produce the isotope ^{211}At. Some cyclotrons with this capability are in operation at universities [NC08] including the University of Washington, University of California at Davis, Duke, and the University of Pennsylvania, though none has cGMP approved hot cells for this purpose. The University of Washington currently produces ^{211}At for a preclinical research program

DOE Higher-Energy Accelerators

Higher-energy DOE accelerators involved in the Office of Science Nuclear Physics isotopes program include the BNL AGS/BLIP and the LANL LANSCE/IPF. These facilities produce "niche" radioisotopes for the DOE that are not available from commercial sources. They also have historically produced research radioisotopes and still produce some of these isotopes for R&D constituencies. Figure 6.1 shows a periodic table of elements indicating which isotopes have been produced.

This list contains both isotopes that require higher energy for production as well as isotopes that can be produced at lower energy but are produced on these machines because of the opportunity cost of parasitic operation. The isotopes that require higher energy for production are limited to ^{82}Sr, ^{67}Cu, ^{28}Mg, ^{32}Si, ^{26}Al, and possibly ^{225}Ac. Because these facilities represent the core of the existing DOE isotope program production capability, they are described in greater detail.

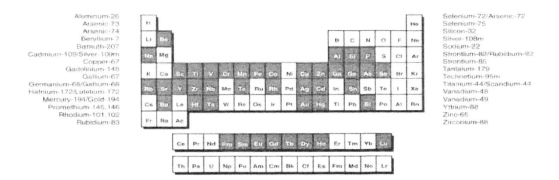

Figure 6.1. Periodic table of the elements showing elements and isotopes historically produced with DOE higher-energy accelerators.

Brookhaven National Laboratory Specific Capabilities

Present Facilities and Capabilities

This program is part of the BNL Medical Department. It uses the *Brookhaven Linac Isotope Producer (BLIP)* and the associated *Medical Department laboratory and hot cell complex* in Building 801 to develop, prepare, and distribute to the nuclear medicine community and industry some radioisotopes that are difficult to produce or not available elsewhere. BLIP, built in 1972, was the world's first facility to utilize high-energy protons for radioisotope production by diverting the excess beam of the 200-MeV proton Linac to special targets. After several upgrades, BLIP remains a world class facility and continues to serve as an international resource for the production of many isotopes crucial to nuclear medicine and generally unavailable elsewhere. The overall effort entails (1) target design, fabrication, and testing; (2) irradiations; (3) radiochemical processing by remote methods in the 9 hot cells of the Target Processing Lab; (4) quality control and analysis; (5) waste disposal; (6) facility maintenance; (7) new isotope and application development; and (8) customer liaison, marketing, packaging, and shipping. Service irradiations (without chemistry) are also performed.

BLIP

BLIP utilizes the beam from the proton Linac injector for the Booster, AGS, and RHIC synchrotrons. The Linac at present accelerates H- ions typically in bunches of 37 mA current, 425 μs duration, and a repetition rate of 6.67 Hz for a time averaged maximum intensity of 105 μA. The beam profile is roughly Gaussian with FWHM of 2.4 cm and 1.8 cm in horizontal and vertical directions respectively. The BLIP generally runs in a parasitic mode, sharing the pulses and operating costs with the driver nuclear physics programs at the Relativistic Heavy Ion Collider (RHIC). The schedule and duration of Linac operation is largely determined by the plans and funding of the nuclear physics experiments, not isotope production needs. The BLIP share and cost depend on the details of the downstream physics programs. If RHIC is running with heavy-ion beams (as, for example, is planned for 2010), parasitic BLIP operation is not available. The average BLIP intensity in parasitic mode is about 20% less than full Linac output, but the cost is 75% less than the full cost. Recently, sales revenue has been sufficient to add operating weeks for BLIP well beyond the RHIC use of the Linac.

Protons of energies of 118, 140, 162, 184 or 202 MeV are diverted down a 30 m long beam line to the shielded isotope production target station. The target assembly is immersed under 9.2 m of water in a 40 cm diameter shaft. The target cooling water is delivered individually past the faces of target disks, then simply empties into the bulk shaft. The height of the water column also provides neutron shielding. There are six mechanically independent target channels, but most recently these have been grouped into two boxes holding up to four targets each. During operations at 118 MeV, the first four targets are sufficient to stop the beam. A hot cell on top of the target shaft is used for target insertion and removal.

Target Processing Laboratory (TPL)

After irradiation, targets are transported in a lead shielded container approximately 0.25 mile to the *Target Processing Laboratory*. This facility contains 9 hot cells. The receiving cell is the largest and consists of one foot thick steel walls with three 18" thick lead glass windows and three master slave manipulators. Targets are cut open inside and the contents transferred to adjacent cells. This hot cell is also used for compacting and packaging radioactive waste (RW). The adjacent cells each have one 14" thick lead glass window, two manipulators and are shielded with 6" lead, clad in 0.5" thick steel panels or equivalent. All cells are equipped with water, reagent addition lines, and RW drains. Aqueous RW flows to three 500 gallon tanks in a basement shielded room for storage. All cells are ventilated through individual roughing filters, then common large charcoal and HEPA filters. In addition, there is a separate acid vapor ventilation system for each cell which neutralizes acid fumes and then joins the main ventilation duct. This is necessary because target dissolution and many other process steps use strong acids and evolve corrosive fumes. The facility maintains a Current Good Manufacturing Practice (cGMP) registration with FDA.

During the 2008 AGS run cycle BLIP successfully delivered isotope products to customers routinely. The BNL isotope program delivered 110 isotope shipments, and isotope sales revenues to DOE were $4.3M in FY08.

Radiochemistry Laboratory

In rooms adjacent to the TPL there are 8 radiochemistry development labs with a total of 14 fume hoods. All have RW drains connected to the waste storage tanks and are exhausted through HEPA filters. Two fume hoods also have charcoal filtration to permit use of radioiodines. Typical equipment including high-performance liquid chromatography, rotary evaporators, centrifuges etc., are available. There is also a manipulator repair area, shipping container storage area, and sizeable staff machine shop.

Instrumentation Laboratory

An instrument room contains three high-resolution gamma spectroscopy systems, two high- efficiency NaI automated gamma counters, a liquid scintillation spectrometer, a UV/VIS spectrometer and an inductively-coupled plasma optical emission spectrometer. There is also a storage cave for counting samples and product retention samples required by

FDA. A second counting room for low level samples contains 4 gamma spectroscopy systems and 4 beta spectrometers.

Training Facilities

This program also hosts one of the Summer Schools in Nuclear and Radiochemistry, sponsored by the American Chemical Society (A second summer school is held at San Jose State University. Each school hosts 12 students each year.). The summer school is an intensive 6 week undergraduate lecture and laboratory course with students competitively selected from all over the U.S. Students receive both lecture and lab course credit from Stony Brook University. This program is very successful, now entering its 19th year. There are two large class rooms and two chemistry labs available in the building, and the students use the low level counting room for the course.

Los Alamos National Laboratory Specific Capabilities

Historical Evolution of Isotope Production at LANSCE

The *Los Alamos Neutron Science Center* is the cornerstone of Los Alamos isotope production. Historically, targets were irradiated at the beam stop with 800 MeV protons at LANSCE from the inception of the facility in the 1970s. The isotopes were produced by a nuclear process known as spallation, which is usually not very selective. It became evident in the mid-1990s that continued delivery of H+ proton beam to the beam stop area would cease because of a lack of programmatic requirements. The isotope program proposed the construction of a new target irradiation facility that would divert beam from the existing H+ beam line in the transition region from the drift tube linac (DTL) to the side-coupled cavity linac (SCCL) into a new beam line and target station housed in a new facility adjacent to the existing accelerator facility. The energy of the protons in this transition region is 100 MeV, and production of isotopes in targets irradiated in this facility occurs primarily by more selective (p,xn) nuclear reactions. Approval was received for this proposal and the construction project was initiated in FY99 and completed in FY03 at a cost of $23.5 M. The 100 MeV Isotope Production Facility (IPF) has operated since the spring of 2004 and irradiates targets while LANSCE is operating for DOE NNSA and Basic Energy Sciences experimental science programs. The 100 MeV IPF has also operated in a dedicated mode when target irradiations from other facilities are not available. An aerial view of the facility is shown in Figure 6.2.

The irradiated targets are transported from LANSCE in a shielded transportation container. The *TA-48 Hot Cell Facility* at the Main Radiochemistry Site, Building RC-1, is the primary hot cell facility for accelerator isotope production. It consists of two banks of 6 chemical processing cells connected at one end by a large multi-purpose "dispensary" cell, where all materials are received, and from which all materials leave the facility. Supporting facilities include several radiochemistry laboratories, a machine shop, two analytical laboratories, an extensive counting room facility, and offices for personnel surround the hot cell facility. This facility, along with the Laboratory's waste handling facilities, is absolutely essential for conducting the LANSCE isotope production mission.

During the latest LANSCE run cycle, the IPF successfully delivered isotope products to customers routinely. The isotope program ships between 150 and 200 isotope shipments annually, and isotope sales revenues to DOE exceeded $4.9M in FY08.

Figure 6.2. An aerial view of the LANSCE accelerator complex with the 100 MeV IPF circled.

Other Los Alamos Facilities and Capabilities

The *CMR Wing-9 Hot Cell Facility* is a complementary hot cell facility to the TA-48 hot cells. The wing-9 hot cells are located in a category 3 nuclear facility and are available for work that requires a nuclear facility safety authorization basis. Currently the isotope program and the weapons program are collaboratively installing a small electromagnetic isotope separator in this facility for the separation of radioactive samples. The facility is also available for the chemical processing of reactor irradiated targets and will be used in conjunction with University reactors, such as the Missouri University Research Reactor (MURR) and the University of California, Davis reactor, to expand the isotope program portfolio of reactor products.

The *TA-50 Radioactive Liquid Waste Treatment Plant* is used to treat and dispose of liquid effluents from isotope production activities. All such effluents are received by TA-50 from an acid waste line that connects both TA-48 and CMR to the facility.

The *TA-54 Solid Radioactive Waste Disposal Site* is used for the storage and permanent disposal of low-level, high-activity waste and low-level, low-activity wastes. All wastes, except mixed wastes, are handled on-site. Currently, the isotope production activities generate no mixed waste, so the isotope program is totally self-contained at the Los Alamos site and is not dependent on off-site facilities for operations.

Development of an isotope production capability proposed as part of the planning, design, and construction of the *Materials Test Station at LANSCE* could represent a significant upgrade to the Laboratory's and the Nation's isotope production capabilities. Production at MTS could complement IPF by re-establishing the LANSCE capability to higher-energy, extending production to other regions of the chart of the nuclides and presenting new and scientifically interesting research and development opportunities for both staff and users. Nuclear reactions of the type (p,xnyp) induced by 800 MeV protons allow the production of a much larger variety of neutron rich isotopes, many of which cannot be produced in a reactor. By irradiating thick targets in the intense proton beam, large quantities of unique isotopes can be produced. In addition to this, a very high flux of both fast and slow neutrons can be generated in the MTS. Some of the irradiation positions available in the neutron flux could be utilized to parasitically produce large amounts of reactor isotopes that are not available commercially.

Examples of Radioisotopes Produced with DOE Accelerators and Isotope Applications

Many of the products available from these high-energy accelerators were developed in collaboration with various research constituencies, and the production and applications R&D was done in parallel. These are examples of the successful production, applications R&D, and successful technology transfer from DOE National Laboratories to the private sector or to the various research constituencies.

- Positron Emission Tomography (PET)
  - ^{82}Sr/^{82}Rb – myocardial imaging
  - ^{68}Ge/^{68}Ga – calibration sources for PET scanners, radiopharmaceutical research
  - ^{72}Se/^{72}As – oncological radiopharmaceuticals
- Isotopes for cancer therapy
  - ^{67}Cu – treatment of non-Hodgkin's Lymphoma
  - ^{103}Pd – seed implants for prostate cancer treatment
  - ^{76}As – bone pain palliation, radiopharmaceutical research for cancer treatment
- Environmental and research radiotracers
  - ^{32}Si – biological oceanography, global climate
  - ^{26}Al – acid rain, Alzheimer's research, materials
  - ^{95m}Tc – technetium behavior in ecosystems

International Accelerator Facilities and DOE Virtual Isotope Center

In addition to DOE Laboratory facilities, the DOE isotope program coordinates the production and output of the DOE supported *Virtual Isotope Center* (Figure 6.3). The concept of a Virtual Isotope Center supported by DOE was born at Los Alamos with collaborations dating back to the early 1990s with TRIUMF in Vancouver, British Columbia, Canada specifically for the production and distribution of the short-lived ^{67}Cu radioisotope, followed closely by collaborations with the Institute of Nuclear Research in Troitsk, Russia. The Virtual Isotope Center has also included the iThemba Laboratory in Cape Town, South Africa, and the Paul Scherrer Institute in Villigen, Switzerland. The beam parameters for each of these facilities are listed in Table 6.2. Originally, the collaborations were directed at ^{82}Sr (See Sidebar 6.1) and ^{67}Cu. Then they were extended to other isotopes as well, including ^{68}Ge, ^{103}Pd, and ^{22}Na. The concept is simple but the logistics are challenging. Targets are irradiated at these international accelerator facilities; they are packaged into appropriate transportation containers and shipped to either BNL or LANL for separation of the isotopes from the target material in the hot cell facility and distribution of the isotopes to DOE customers. Even with the difference in the beam parameters listed in the Table 6.2, the isotope products all meet DOE and customer specifications.

Current Status and Impacts of the Production Capability

The current state of the production capacity is captured in the bulleted information below. They describe the major products produced in each type of accelerator and provide the impact that each production capability has on isotope availability.

- Commercial cyclotrons: Currently produce ^{201}Tl, 111In, ^{123}I, and other commercial isotope products. These cyclotrons satisfy customer requirements for commercial accelerator isotopes but have no impact on research isotope availability.
- University PET cyclotrons: Currently produce ^{18}F, 11C, ^{15}O for PET. Currently, most are underutilized with respect to beam availability. These cyclotrons satisfy needs of clinical PET centers and PET research programs at each University but have a marginal effect on other research isotope availability. They could have a much larger impact if efforts for coordinated production from these facilities could be realized.
- DOE higher-energy accelerators: Currently produce ^{82}Sr, ^{68}Ge, ^{22}Na, ^{73}As, and are capable of enhanced research isotope production. Each facility has extensive hot cell facilities to support isotope production. These accelerators and complementary facilities satisfy customer requirements for "niche" commercial accelerator isotopes and have some impact on research isotope availability. They could make a much larger impact with additional funding resources and additional beam availability but are not cost-effective to operate for isotopes that can be produced at lower energies.
- International high-energy accelerators: Currently produce isotopes independently and in collaboration with the DOE. Processing capabilities are variable from accelerator to accelerator. These collaborations help to satisfy customer requirements for "niche" commercial accelerator isotopes and have local impacts on research isotope availability. These accelerators could also have a greater impact depending on resources.

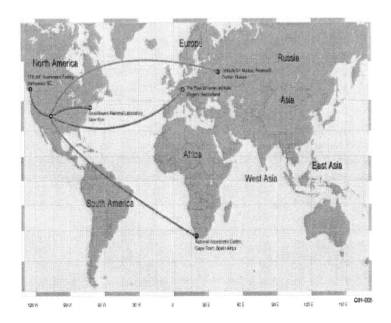

Figure 6.3. DOE Virtual Isotope Center concept.

Current Deficiencies in the Production Capability

Clearly from the discussions above, the major deficiencies in the current accelerator isotope production capability are 1) in using facilities whose primary missions are not isotope production, it is difficult to guarantee the availability and variety of isotopes needed by multiple research constituencies and 2) for isotopes better produced at lower energies, the current isotope program facilities are not cost efficient. There is a real need for new capability that is optimally configured for research isotope production. Ideally, such a new capability would include a dedicated facility whose primary mission was isotope production, with secondary missions of isotope production R&D, technology development and transfer, and education and training and the associated infrastructure required to pursue these multiple missions. The facilities needed would include an irradiation facility; hot cell processing facilities; waste handling and disposal facilities; radiochemistry laboratories for R&D; support laboratories for radiometric assays, target fabrication, and quality assurance; and associated facilities for the education and training mission. Conduct of the education and training mission and the governance model for such a dedicated facility would be dependent on the specifics of the site selected for the dedicated capability.

Table 6.2. Comparison of higher-energy accelerator facility beam parameters

Parameter	IPF	BNL	INR	iThemba	TRIUMF	PSI
Beam energy (MeV)	100	200	160	66	500	72
Beam current (μA)	250	105	100	120	150	100
Beam time structure	13.3 mA 625 μs 30 Hz	39 mA 420 μs 7 Hz	27 mA 74 μs 50 Hz	DC	DC	DC
Beam spot (FWHM in mm)	20 swept	19 × 12.5	8 - 10	5 swept	32	
Peak power Density per MeV (in W/cm^2)	80 26 swept	64	127 56 slanted	509 48 swept		
Target diameter (mm)	50	75	25 at 26°	20	75	
Cooling channel width (mm)	5	5	3	1	2	
Cooling water velocity (m/s)	2	1.5	high	20		

Isotopes for the Nation's Future: A Long Range Plan

For reasons discussed in the sections above, when focusing on research isotopes the mission is best served by a 40 MeV cyclotron with multiple particle capability. Expected beam energies for a 40 MeV variable-energy, multi-particle machine include

- protons 20-40 MeV variable,
- deuterons 10-20 MeV variable, and
- alphas 40 MeV (positive ion).

Sidebar 6.1: ^{82}Sr: A Case History for International Cooperation.
An Isotope's Journey from Research Application to Clinical Use

Strontium-82 is the parent isotope of ^{82}Rb, and this isotope pair, ^{82}Sr/^{82}Rb, is the basis for the CardioGen ^{TM}biomedical generator marketed by Bracco Diagnostics, manufactured for Bracco by GE Healthcare, and used in clinical Positron Emission Tomography for cardiac perfusion studies. The production of ^{82}Sr requires higher energy (70 MeV) and the original production of this research isotope was pioneered by the DOE high-energy accelerators. Both BNL and LANL also performed R&D on the development of the biomedical generator concepts. From the late 1970s until the biomedical generator was approved by the FDA for clinical use in 1991, both BNL and LANL produced the isotope for the development work. During this period the Canadian firm, Nordion (now MDS Nordion) also developed the capability to produce this isotope. When the generator became commercial in 1991, it was important that all three ^{82}Sr suppliers remain in the mix because the high-energy accelerator schedules are such that all three suppliers are required to have year round availability of the ^{82}Sr.

Even with three suppliers it became evident in the summer of 1997 that there would be supply problems from January through May of 1998 because all three accelerators were going to be down simultaneously. Fortuitously, LANL had been collaborating with the Institute of Nuclear Research in Troitsk, Russia, under the auspices of the Initiatives for Proliferation Prevention Program of the DOE, and one of the R&D focus areas was developing a joint capability (INR doing target irradiations and LANL doing chemical processing) for ^{82}Sr since 1996. In the four months leading up to the time when the first targets needed to be irradiated to meet the January 1998 delivery dates, LANL and INR resolved all of the logistics issues associated with the joint production capability (transportation of irradiated targets, customs/export control, FDA approval, etc.) and received FDA approval to meet the delivery dates in 1998 with Russian-irradiated rubidium metal targets. Subsequent to this time, the DOE Virtual Isotope Center was expanded to include iThemba Labs, in Faurve, South Africa, and the Paul Scherrer Institute (PSI) in Villagen, Switzerland. Between the international collaborations, the DOE National Laboratories, and the MDS Nordion production at TRIUMF, there has never been a missed delivery of the ^{82}Sr isotope to the generator manufacturer.

The history of ^{82}Sr is both a case history in the development time for a research isotope to become a commercial product, and the value of national and international collaborations and cooperation in insuring the year-round availability of short-lived radioisotopes.

The major benefits of a 40 MeV machine include the following:

- Control of beam energy in the low-energy range is much more accurate due to the reduced effect of energy straggling and multiple scattering. This is the major nuclear physics based benefit over a 70 MeV machine.

- Using targetry with 2, cooling (standard plating on water cooled backing used for 30 MeV machines): (p,n), (p,2n), and (p,3n) reactions can be utilized. The tunable primary beam energy down to 20 MeV allows the avoidance of impurities produced by the (p,3n) reaction.
- Using targetry with 4, cooling (cooling water layer in front of the target): The extra 10 MeV over a commercial 30 MeV machine makes this type of targetry possible. This arrangement also allows the use of encapsulated targets which opens up a very large body of possible target materials. Encapsulated targets in a 4, cooling arrangement can still utilize both (p,n) and (p,2n) reactions. About 10 MeV is lost in the beam window, cooling water layer, and the encapsulation, allowing 30 MeV protons to reach the target material.
- The targetry flexibility and the enhanced ability to conduct nuclear data research and development relevant to isotope production are major advantages over higher-energy machines.
- The cost of facilities for lower-energy machines and the on-going operating costs are advantageous compared to higher-energy alternatives. While the cost benefits do not scale linearly with energy, there are large cost savings over the life cycle of the irradiation facilities.
- Lastly, development of isotopes and production methodologies at these lower energies are more directly transferable to industry when the isotopes become commercially viable, as has been demonstrated for the last 50 years with 30 MeV cyclotrons. Because of capital costs and operating costs, this is less feasible or likely with higher-energy machines.

Examining the list of requirements for research isotopes for the next five years generated by the NIH/DOE working group, such a facility would address the accelerator-produced isotopes whose availabilities were in question, ^{211}At, ^{76}Br and ^{77}Br. It is well understand that research priorities can and will change, but this re-enforces the subcommittee's position that the isotope program needs capability to produce research isotopes at both low and high energy.

The subcommittee considered the merits of a higher-energy machine at 70 MeV. This choice would allow for more cost effective production of most of the isotopes currently produced at IPF and BLIP. However, these variable energy cyclotrons typically can only extract high quality beams from 50-100% of the maximum energy and alpha beams at the maximum energy. As Table 6.1 shows, most of the research isotopes are best produced at lower energy, and the subcommittee places priority on effectively producing a variety of research isotopes, not just a few. One important example is ^{211}At, where the alpha beam energy needed in the production target is about 26-28 MeV, not 70 MeV. If a higher-energy, multi-particle, variable energy accelerator were available that could extract high quality beams from 15 to 70 MeV, this option would be an excellent choice for the new capability. Such designs do not appear to be available commercially at present. On the other hand, if the current parasitic operation of BLIP or IPF were to become unfeasible in the future due to termination of the primary DOE missions for these facilities, a higher-energy accelerator to more cost-effectively replace these capabilities must be considered.

Scientific and Technical Challenges

The scientific and technical challenges of accelerator isotope production and applications R&D are myriad. Any new isotope production methodology for a research isotope involves science challenges that include everything from nuclear physics to materials science to chemical separations technology to product quality and quantity to waste identity, handling and disposal. In addition, non-traditional methods of production are also exciting R&D possibilities. The bulleted information below provides a list of the opportunities.

- New research isotope production at any energy requires R&D in the following areas:
  - o Nuclear reaction cross sections/nuclear physics, including maximizing desired isotope yields and minimizing impurity production.
  - o Targetry development, including materials science (compatibility, material behavior, etc.), thermal hydraulic modeling, and chemistry (target dissolution).
  - o Processing chemistry, including chemical separations, waste minimization, and waste characterization, handling, and disposal.
  - o R&D activities to determine product quantity and quality, and in many cases, determine whether a particular product is suitable for applications R&D.
- Research into alternative isotope production technologies to complement traditional techniques, including alternate accelerator technologies and isotope separator technology for targets.

Most Compelling Opportunities and Impacts

After evaluating the existing capabilities and determining the contributions of these capabilities, the following opportunities and impacts have been identified. Some are steps that can be taken with minimal new financial resources and may have an impact on the necessity and magnitude of new capital investments. However, even with these improvements, the consensus of the subcommittee is that additional capital investments will be required to augment the incremental improvements that can be made by better utilization of existing capabilities.

- Better coordination and utilization of beam availability at university PET cyclotrons, nuclear pharmacy/hospital cyclotrons, and DOE accelerator facilities to enhance and expand the research isotope portfolios from these non-DOE facilities. Some efforts in these directions have been attempted, but they should be dramatically expanded.
- More extensive collaborations with international high-energy accelerator facilities to supplement and enhance DOE's existing production capabilities.
- Dedicated accelerator facilities for isotope production.
- Non-traditional accelerator approaches to isotope production (e.g., electron accelerators).

Relationships of Existing and Future Capabilities

The following relationships have to be considered in future planning about additional capital investments versus better utilization of existing capabilities. Even with these considerations and the better utilization of relationships, the committee has determined that additional capital investments are required, and these are included in the priority recommendations.

- Existing Capabilities
  - o DOE higher-energy accelerators at BNL and LANL already coordinate schedules to extend availability. Cooperation should be extended to research isotopes as resources are available.
  - o DOE's "Virtual Isotope Center" already utilizes international high-energy accelerators, but this cooperation can be extended as required for research isotopes.
  - o There are significant ongoing maintenance and infrastructure needs required to preserve the current production capabilities. As discussed in Chapter 11, resources to address these needs have been very limited in recent years.
  - o A core of specialized, highly trained personnel is essential to operate and improve these facilities and to develop new isotope production techniques. As discussed in Chapter 11, a significant number of these individuals are supported by sales revenue, and fluctuations in sales can put the availability of these individuals in jeopardy.

- Enhanced Utilization of Existing Capabilities

  - o Additional beam time and additional processing resources directed toward the production of research isotopes.
  - o Incorporation of Arronax and other such facilities that may come on-line into the "Virtual Isotope Center" and an increase in target irradiations for commercial isotopes so DOE accelerator beam time can be redirected to research isotopes.

- Utilization of Untapped Capabilities

  - o Utilization of unused beam time at university PET cyclotrons. An open question is the use of DOE processing capacity versus establishing small scale processing capability at a select number of cyclotron facilities.
  - o Explore the utilization of commercial cyclotron unused beam time if there is any.

- Future Capabilities

  - o Future accelerator capabilities should be used to complement existing capabilities with some exceptions. The major priority for production of accelerator research isotopes should be a lower-energy (40 MeV), multiple particle (hydrogen ions, alphas) cyclotron. Such a facility fills a huge gap in research isotope accelerator production capability both domestically and internationally. This cyclotron will be energy tunable down into the range of University PET cyclotron (~1 5 MeV) and will not only insure research isotope availability for 90+ % of the desired isotopes but also allow for the isotope production research (e.g., cross section measurements) necessary to make the research isotopes available.
  - o Incorporate new private/public sector initiatives (e.g., Duke's planned Biological Accelerator Complex facility).
  - o Explore alternate accelerator technologies.

Research Isotope Availability

The availability of research isotopes has been an on-going issue for various domestic and international research constituencies for many years. A thorough examination of the use of isotopes in commercial products and in established R&D efforts is illustrative of this lack of availability. Most of the isotopes in use today in practical settings were developed as long as 50 years ago. With few exceptions (e.g., ^{82}Sr and ^{90}Y) there are no new products or services that use isotopes developed in the past 20 years. Without the availability of research isotopes, it is not possible to develop new science or new applications based on isotopes. This problem is extreme in the case of accelerator isotopes and less extreme for reactor isotopes because of the efforts and availability of MURR and HFIR. The problem is exacerbated for accelerator isotopes because of the parasitic nature of operations of the DOE higher-energy accelerators, the lack of established techniques to increase utilization of university and nuclear pharmacy/hospital cyclotrons (which could be modeled after the Washington University NIH/NCI efforts), and the lack of facilities dedicated to research isotope production. The problems that have led to the lack of research isotope availability also foreshadow the solutions, some of which are listed below.

- Increased funding for DOE higher-energy accelerators, both operations (including increased beam time) and infrastructure, can immediately ameliorate the situation but is only a small part of the solution to the problem. Capacity gains will be dependent on additional available beam time.
- Increased coordination and utilization of university cyclotrons will require funding for beam time and logistical costs associated with matching cyclotron irradiations with appropriate processing facilities. To fully develop this scenario, a production planning exercise against an availability scenario will have to be completed.
- New facilities must include dedicated accelerator facilities and may also include radioactive material isotope separator facilities and, possibly, small-scale hot cell facilities for selected university and nuclear pharmacy/hospital cyclotrons.

Recommendations

Based on these considerations, the subcommittee makes the following recommendations related to accelerator produced isotopes:

Program Operations

Maintain a continuous dialogue with all interested federal agencies and commercial isotope customers to forecast and match realistic isotope demand and achievable production capabilities.

Coordinate production capabilities and supporting research to facilitate networking among existing DOE, commercial, and academic facilities.

Support a sustained research program in the base budget to enhance the capabilities of the isotope program in the production and supply of isotopes generated from accelerators.

Increase the robustness and agility of isotope transportation both nationally and internationally.

Highly Trained Workforce for the Future

Invest in workforce development in a multipronged approach, reaching out to students, post-doctoral fellows, and faculty through professional training, curriculum development, and meeting/workshop participation.

Major Investments in Production Capability

Construct and operate a variable-energy, high-current, multi-particle accelerator and supporting facilities that have the primary mission of isotope production.

The most cost-effective option to position the isotope program to ensure the continuous access to many of the radioactive isotopes required is for the program to operate a dedicated accelerator facility. Given the uncertainties in future demand, this facility should be capable of producing the broadest range of interesting isotopes. Based on the research and medical opportunities considered by the subcommittee, a 3 0-40 MeV maximum energy, variable energy, high-current, multi-particle cyclotron seems to be the best choice on which to base such a facility.

7. REACTOR BASED ISOTOPE CAPABILITIES

Research reactors have historically played and will be expected in the foreseeable future to continue to play a critical role in the production of isotopes used for research and commercial applications. Reactors operate by a chain reaction based on the neutrons released in fission of, most commonly, the isotope ^{235}U. Therefore, they are powerful continuous sources of neutrons. As neutrons have no electrical charge, they can penetrate into the atomic nucleus of a target even at very low energy, and the probability of nuclear reactions can be much larger (in some cases, several orders of magnitude) than is the case for charged particle beams. This same characteristic that enables a reactor to work at all makes reactors particularly powerful sources either for isotopes where neutrons are added to the target or for isotopes produced by neutroninduced-fission.

The peacetime production of radioisotopes at reactors began in 1946 at the Graphite Reactor in Oak Ridge, TN, under the management of Waldo Cohn of Clinton Laboratories and Paul Aebersold of the Atomic Energy Commission. In August of that same year the Laboratory's research director, Eugene Wigner, handed the first shipment of reactor produced radioisotopes, a container of carbon-14, to the director of the Barnard Free Skin and Cancer Hospital of St. Louis, MO. Over the next two decades, the use of reactor-produced isotopes permeated nearly every field of science (Sidebar 7.1).

Sodium-24 was one of the first radionuclides used to measure the permeability of canine red blood cells *in vivo*. The tracer approach was quickly applied to clinical situations such as the study of thyroid metabolism using radioactive iodine and uptake, and the retention and excretion of radiolabeled phosphorus. These studies provided valuable information about the selectivity of proposed therapeutic regimens [IM95].

In addition to environmental and clinical applications, physicists quickly realized that the promise of nuclear power required a thorough understanding of isotopes produced in the fission process and their effects on the efficiency and safety of nuclear power generation and, ultimately, on the safe disposal of spent fuel.

Commercial applications also developed over this time period. The use of californium-252 as a neutron source for radiography quickly became an important technique to understand material stressors and predict failure events through nondestructive evaluation.

The U.S. government's active production and promotion of the use of radioisotopes for research and commercial purposes also led to increased demand for these materials. In a sense, this engagement by the government demonstrated U.S. commitment to harness the atom for peace. The scientists of the world responded (Figure 7.1).

Figure 7.1. Oak Ridge National Laboratory's Isotope Circle pictured in *LIFE* Magazine.

Landscape of Reactors

Worldwide today, 278 research reactors are known to be operating in 56 countries supporting a variety of tests, training, and research missions including isotope production. The majority of these reactors are over 30 years old, and the number of shutdown or decommissioned research reactors is around 480, by far more than are operating. In the United States, there are currently 32 operating research reactors that are licensed by the U.S. Nuclear Regulatory Commission (NRC) and two of the four reactors operated by the U.S. Department of Energy are used for research and isotope production (Figure 7.2). Most of the U.S. reactors are over 40 years old; however, many have recently completed or are currently in the process of being relicensed for an additional 20 years. There are 18 U.S. research reactors with power levels greater than one megawatt (thermal). Three of these reactors are uniquely suited for the production of isotopes: (1) the University of Missouri Research Reactor (MURR), (2) the High Flux Isotope Reactor (HFIR) at Oak Ridge National Laboratory, and (3) the Advanced Test Reactor (ATR) at Idaho National Laboratory. Each of these reactors has sufficient flux and in-core irradiations facilities to allow for the production of isotopes that generally require a reactor production method.

The University of Missouri Research Reactor (MURR) operates at 10-megawatts (thermal) with a peak flux of 6×10^{14} n/(cm^2 sec) and is the most powerful research reactor located on a U.S. university campus. MURR features multiple irradiation facilities covering a spectrum of neutron fluxes and geometries. MURR's weekly operating cycle makes it a key supplier of a broad range of radioisotopes for research, education, and industry. The reactor is at full-power operation 52 weeks per year. An example of MURR's bulk isotope product development is the production of two radioisotopes of phosphorus ^{32}P and ^{33}P, now used by researchers throughout the world in protein and DNA analysis. The radiopharmaceutical research group at MURR focuses on development of radioisotopes for use in detecting and treating cancer and other chronic human diseases. Additionally, MURR is now engaged in an initiative to become a domestic supplier of Molybdenum-99 (^{99}Mo). As discussed more extensively in Sidebars 4.3, ^{99}Mo is the parent isotope of technetium-99m, which is used for medical diagnostic procedures performed approximately 35,000 times daily in the United States alone. MURR's strategic objective is to supply up to half of the U.S. demand for ^{99}Mo.

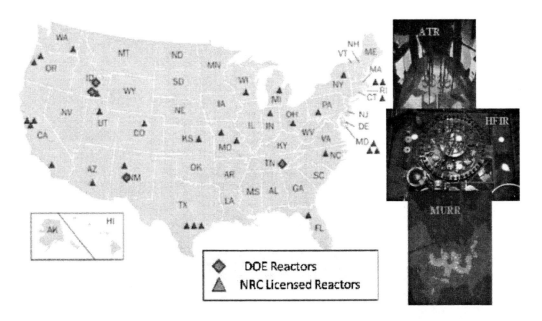

Figure 7.2. Reactors of power greater than 1 MW in the U.S.

At 85-megawatts, the Oak Ridge National Laboratory High Flux Isotope Reactor (HFIR) provides one of the highest steady-state neutron fluxes of any research reactor in the world. HFIR's primary mission for the DOE Office of Basic Energy Sciences is for neutron scattering research and materials studies. HFIR provides a peak flux of 2.6×10^{15} n/(cm^2 sec) and operates up to nine 23-day cycles per year.

Sidebar 7.1: Reactor Radioisotopes and Health

"If at some time a heavenly angel should ask what the laboratory in the hills of East Tennessee did to enlarge man's life and make it better, I daresay the production of radioisotopes for scientific research and medical treatment will surely rate as a candidate for the very first place."
- Alvin Weinberg

The peacetime production of radioisotopes at the Oak Ridge National Laboratory (ORNL) Graphite Reactor for industrial, agricultural, and research applications began in 1946 under the management of Waldo Cohn of Clinton Laboratories and Paul Aebersold of the Atomic Energy Commission (AEC).

In August 1946, the Laboratory's research director, Eugene Wigner, handed the first shipment of a reactorproduced radioisotope, a container of carbon-14, to the director of the Barnard Free Skin and Cancer Hospital of St. Louis, Missiouri

During its first year of production, the Laboratory made more than 1000 shipments of 60 different radioisotopes, chiefly Iodine-131, Phosphorus-32, and Carbon-14. These were used for cancer treatment in the developing field of nuclear medicine and as tracers for academic, industrial, and agricultural research. Many thousands of shipments of radioisotopes produced at the Graphite Reactor were made before production was shut down permanently in 1963.

Following the closing of the Graphite Reactor, the Oak Ridge Research Reactor produced most of the Laboratory's radioisotopes. The Oak Ridge Research Reactor closed in 1987, but ORNL's High Flux Isotope Reactor remains an important source of radioisotopes for medical and industrial uses. The Laboratory's nuclear medicine program now centers on the development of new radiopharmaceuticals and radionuclide generators for diagnosis and treatment of human diseases, including cancer and heart ailments.

Originally designed to produce usable quantities of heavy actinide isotopes, HFIR remains the sole producer of Californium-252 in the Western hemisphere (used in a broad range of applications: cancer treatment, non-destructive evaluation of aircraft components, and detector technologies to protect U.S. borders to name a few (See Sidebar 4.2). The HFIR is also equipped with a rabbit facility enabling the injection, irradiation, and removal of material in its core during operations. HFIR produces 35 primary isotopes including Palladium-103 (for aid in the treatment of prostate cancer), Rhenium-188 (for the treatment of cancer and arthritis), Iridium-192 (for cancer treatment therapy to reduce metastasis, for preventative treatment of post vascular surgery complications related to angioplasty, for oil well exploration and for the conduct of geological surveys and radiographic inspection of components) and Tin-117m (for palliative bone cancer treatment). The operating schedule for HFIR lends itself to research-quantity, high-specific-activity isotopes and to production of the heavy actinides and longer half-life isotopes. However, because of its running schedule and other mission requirements, it is not well suited for large-scale production of short half-life radioisotopes such as ^{99}Mo.

The Idaho National Laboratory Advanced Test Reactor (ATR) is designed primarily as a fuels and materials test reactor for the DOE/NNSA Naval Reactors program and the DOE Office of Nuclear Energy. Because of its high flux and large volume of irradiation space, the ATR lends itself to isotope production as well. ATR can operate at 250 megawatts with a maximum flux of 1×10^{15} n/(cm^2 sec), although the constraints of its materials irradiation mission typically require it to operate at lower levels. Currently, ATR operates about five cycles per year with each cycle lasting approximately 50 days, depending on the power levels required by its experimenters. Although not a primary mission, ATR is suited to produce quantities of high specific activity isotopes for medical and industrial applications. One such isotope is Cobalt-60 which is used in a medical device known as the "gamma knife" that provides precise treatment of otherwise inoperable vascular deformities and brain tumors. Also, ATR has been designated as the principle site for the future production of Plutonium-238 used in radioisotope thermoelectric generators (RTGs). RTGs are used to produce reliable power over longer periods of time and can be subjected to extreme environments (Sidebar 3.B.3). They are a principal power source utilized by the National Aeronautics and Space Administration (NASA) aboard deep-space exploration vehicles. INL is investigating strategies such as rabbit target transport systems to be able to remotely insert, radiate, and remove isotope production targets during a run cycle.

The smaller research reactors also provide important radioisotopes to researchers and commercial radiotracer companies. These users generally have a need for small batch quantities of relatively short-lived radioisotopes.

Utilization

During the course of this review, approximately 40 reactor-produced isotopes were identified in a recent Nuclear Materials Assessment Report of the Isotope Business Office (See Table 7.1) that have been requested by the commercial and scientific communities, but have not been made available by the isotope program. In all cases, the unavailability of the isotope was NOT the result of there being no irradiation facility (reactor) available to produce the material. In general, the United States has sufficient research reactors to meet research isotope production demands. The fact that many requested isotopes are not being produced can be attributed more to the high cost (especially for small batch production) which cannot be realized with current funding and a lack of programmatic coordination among the user community and the production facilities. Better utilization of existing research reactor assets in the United States is necessary to meet current and anticipated isotope demands. However, it is believed that the existing capacity is sufficient.

The ability of the program to predict demand for certain isotopes needs vast improvement. The only way to understand what isotope will be requested in the future is to engage the community. Isotope production often requires one or more years of activity before the final isotope product can be made available to the user. This lead time is not well understood by the user community and can be a barrier to obtaining the funding necessary to do research requiring a particular isotope.

In order to be in a position to respond efficiently and in a timely manner to production requests, programmatic base funding to maintain target fabrication, processing, and source fabrication (Figure 7.3) is necessary. In order to meet demands, the facilities and expertise must stay "mission ready." More regular support for facility maintenance and infrastructure issues is needed. As discussed for other isotope production capabilities, a unique core of highly specialized personnel is required for this mission-readiness. At present, too many of these individuals are funded directly from isotope sales, with the resulting uncertainty due to demand and sales fluctuations.

Table 7.1. Reactor-Produced Isotopes (from No Materials Available List)
($\sqrt{}$ = notional prioritization)
Production Options: HFIR, ATR, MURR, and Other (university research reactors)

Isotope	Production Option	Comments
Ac-227 $\sqrt{}$	HFIR	REDC processing advantage
Ba-133 $\sqrt{}$	HFIR, ATR	
Co-60 LSA	HFIR, ATR	Have necessary reactor volume
Co-60 HSA	HFIR	Flux & volume advantage
Cm Isotopes	HFIR	Associated w/Cf prod, REDC processing
Gd-153	HFIR, ATR, MURR	High gamma waste produced during process
Isotope	Production Option	Comments
Ir-192	HFIR, ATR	Source makers want US supply
Np-237	HFIR	U-238 target material
Pm-147$\sqrt{}$	ORNL	Fission product or indirect from Nd-147
Pu-238	ORNL	REDC processing advantage
Pu-244	ORNL	REDC processing advantage
Re-186 HSA	ORNL	
Ru-106	ORNL	Fission product
Sr-90	ORNL, ATR	Fission product. Low demand.
Th-228	ORNL	REDC processing advantage
Fe-59	MURR, other	
HG-203	MURR, other	

Table 7.1. (Continued)

Isotope	Production Option	Comments
Mn-54	MURR, other	
Nb-94	MURR, other	
Nd-147	MURR, other	
Ni-57√	MURR, other	
Ni-65√	MURR, other	
P-32√	MURR, other	
Pb-200	MURR, other	
Pb-202	MURR, other	
Pb-203	MURR, other	
Pd-103√	MURR, other	
Pr-142√	MURR, other	
Sb-125√	MURR, other	
Sc-44√	MURR, other	
Se-72√	MURR, other	
Se-75	MURR, other	
Si-32	MURR, other	
Sm-151	MURR, other	
Sm-153√	MURR, other	
Sr-89	MURR, other	
Tc-97	MURR, other	
Tl-204 √	MURR, other	
Tm-171√	MURR, other	
Yb-169	MURR, other	

However, better utilization of existing reactor facilities could improve program performance. In some cases the reactor facility could be improved through a better understanding of the in-core flux characteristics, a better understanding of nuclear cross sections with some target material, or improved in-core isotope production facilities such as improved flux trap or "rabbit" designs.

In addition, supply reliability could be dramatically improved through increased coordination and collaboration among the production facilities. In these considerations, reactor production facilities are not different from the accelerator production facilities. Each should operate collaboratively as a system of facilities instead of individual, often competing, elements. To improve collaboration, dramatic improvements in the ability to transport isotopes in various forms between production sites and, eventually, to the end user are needed. The subcommittee identified a real and pressing need to certify new transportation containers for a variety of research isotopes (See Chapter 11). In particular, a new flexible type B container is urgently needed. This design should include features such as high activity gamma shielding that is compatible with production facility hot cell and handling equipment, a draining capability that allows for loading directly from pool type reactors, and incorporation of "next generation" security features. There is also a need to produce both type

A and type B containers (defined based on the level of radioactivity of the isotope being shipped) that are light weight, allowing for more flexible carrier selection (over the road or air carrier), and that can handle isotopes in solid or liquid form. Examples of transportation packaging and infrastructure are shown in Figure 7.4. Improvements in the transportation of isotopes would greatly enhance the program's ability to operate production facilities in a collaborative manner.

Figure 7.3. Target assembly and fabrication examples.

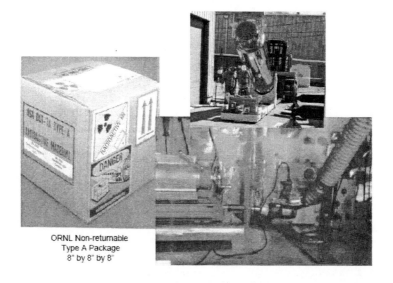

Figure 7.4. Examples of transportation packaging and infrastructure.

International Cooperation

Isotope supply would certainly be improved if isotopes produced around the globe were standardized. Many of the isotopes produced in the United States are supporting researchers around the world, and the U.S. research community is dependent upon overseas suppliers for many isotopes used in the United States. The global nature of isotope production and use was never more evident than it was this year when the world's ^{99}Mo supply was disrupted after a series of safety related shutdowns of reactor production facilities (See Sidebar 4.3).

Sidebar 7.2: Reactors to Produce Neutron-Rich Isotopes

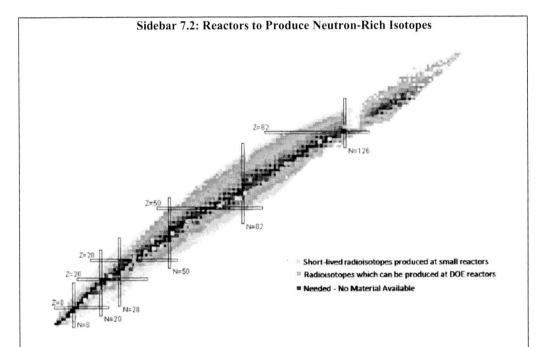

From the time of the earliest design of nuclear reactors it has been recognized that a major application of a reactor would be for the production of radioisotopes. In general, nuclear reactors are useful for the production of neutron-rich isotopes, that is, nuclides to the right of the curve of stability. Because of this characteristic, reactors are most useful for producing radioisotopes that decay by alpha and beta particle emission, the predominant decay modes of neutron-rich radioisotopes.

An important parameter that must frequently be considered in reactor production of radioisotopes is the specific activity of the product. For direct production of radioisotopes (i.e., the target and product are isotopes of the same element), specific activity is a function of the reactor neutron flux (magnitude and energy distribution) and the cross section of the target material as a function of neutron energy; in these cases, greater neutron flux produces higher specific activity products. Another important reactor design consideration is the volume available for irradiation which will be a factor in determining the size of the irradiation target and, ultimately, the total amount of radioisotope produced. In general, to produce large quantities of high-specific-activity radioisotopes, reactors of high flux and large available irradiation volumes are desired. High neutron flux is also particularly important when the isotope being produced requires multiple neutron captures since flux to the power of the number of neutron captures is a dominant term in the equation for isotope production rate.

In this case, the United States had no domestic capability to bridge the supply gap. For many isotopes there may only be one or two production sources in the world. Californium-252 is another good example of this situation (See Sidebar 4.2). Currently only two facilities in the world have the ability to produce this heavily used isotope. If production at one of these facilities is disrupted for any appreciable period of time, this material will simply be unavailable. It is for these reasons that the isotope production program must work to improve international collaboration.

The International Atomic Energy Agency (IAEA) has recognized this need, which is more acute in countries that have no domestic production capability and must rely on their neighbors or other foreign sources for research isotopes. In response, the IAEA has established regional coalitions of research reactors. It is their opinion that there exist a large number of research reactors in the international community that are not fully utilized for large-scale radioisotope production. Due to various reasons such as lack of funds and trained manpower, they are unable to advertise their capabilities or elicit the support of end users. This underutilization could result in prolonged planned or unplanned shutdowns, fundamentally affecting the sustainability of the reactor itself.

To resolve this issue internationally, attempts are being made to develop regional strategies for the effective and efficient utilization of these reactors taking into account national as well as regional needs. The IAEA is working within the international community to improve collaboration and, thus, improve utilization. (Technical Working Group on Research Reactors Meeting, March 2009, Vienna, Austria). The approach is an international expansion of the philosophy the subcommittee envisions for the United States, that is planning isotope production systematically considering all domestic production facilities, equipment and intellectual assets. The NSACI subcommittee supports this effort.

Planning for the Future

Although this subcommittee has concluded that existing research reactors have the capability and capacity to meet almost all (^{99}Mo is perhaps the notable exception) current and projected isotope production demands, there is a very real limitation in how long the United States can count on them. As stated earlier, most research reactor assets are on average 40 years old. It is true many have extended their operating horizon by another twenty years, and in some cases, believe that their reactor can operate as long as eighty years. It is unreasonable to believe that they can safely operate beyond this. The MURR reactor will be 43 years old this year, HFIR has operated 44 plus years, and ATR is approaching its 42nd birthday. Long-range planning needs to begin now to develop a strategy for reactor produced isotopes beyond the life of these facilities (Sidebar 7.3). New research reactors take decades of planning to build and make operational. Worldwide there are 8 new reactors under design or construction, most of which are not being built specifically for isotope production. There are currently no plans for replacement capability in the United States for reactor produced isotopes.

Recommendations

Program Operations

Maintain a continuous dialogue with all interested federal agencies and commercial isotope customers to forecast and match realistic isotope demand and achievable production capabilities.

Coordinate production capabilities and supporting research to facilitate networking among existing DOE, commercial and academic facilities.

Support a sustained research program in the base budget to enhance the capabilities of the isotope program in the production and supply of isotopes generated from reactors.

Increase the robustness and agility of isotope transportation both nationally and internationally.

Highly Trained Workforce for the Future

Invest in workforce development in a multipronged approach, reaching out to students, post-doctoral fellows, and faculty through professional training, curriculum development, and meeting/workshop participation.

**Sidebar 7.3. Future Research Reactors for
Isotope Production and Materials Irradiation**

As mentioned in this report, most research reactor assets are on average 40+ years old. Long- range planning of the next generation of research reactors should be taking place now so they will be ready to begin operating as the current reactors near the end of their life. This process will involve years of planning and construction in order to accomplish a smooth transition.

A team of ORNL researchers brainstormed this subject and developed what they think are the characteristics of the next generation reactor. This team included research staff having expertise in (1) isotope production, (2) experiment design for the High Flux Isotope Reactor (HFIR) and the Advanced Test Reactor (ATR), (3) the science of irradiated materials, and (4) reactor design/licensing. The characteristics presented below represent their consensus of the attributes that a world class research reactor facility should strive to attain. The team further determined that a reactor could be designed which would meet most if not all of these characteristics. Flexibility, simplicity, ease of operations including loading and unloading of targets, and variable neutron energy spectrum were considered essential. Summarized below are the reactor and experimental facility characteristics that would optimize isotope production and materials irradiation and are considered important for the next generation of reactors. These characteristics are described below in no particular prioritized order.

General Reactor Characteristics

The next generation reactor should be expected to use low enriched uranium (LEU) fuel and light water coolant. A swimming pool concept similar to that used in the HFIR and the Oak Ridge Research Reactor (ORR) is preferred because it allows for easy access to the core and experimental locations. The core should have adequate space to host a number of user facilities. It was suggested that the fuel form should be more standardized such as the box-type fuel elements used in ORR. The reactor should have an array of many fuel and reflector elements as opposed to a fixed fuel arrangement utilized at HFIR which has only two annular elements in a fixed configuration. This design allows more flexibility, ease of fabrication, and increased availability.

Experimental Facility Characteristics

The reactor facility should have a hot cell that has access to the reactor pool. This capability allows targets to be transferred without the need for several large shipping containers and greatly enhances the ability to quickly encapsulate and/or de-encapsulate targets and assemblies removed from the reactor. The reactor should be equipped with at least 3 hydraulic rabbit facilities, each with adequate cooling to accommodate high thermal loads. A Neutron Activation Analysis pneumatic rabbit facility similar to the current system at HFIR should also be included in the new reactor, including a low background counting room.

It is important to maintain the ability to produce transplutonium isotopes in some portion of the core. Given the requirement of using LEU fuel, a high density fuel such as U-Mo would be required to achieve this goal.

The box-fuel array core design should allow open loop experimental irradiation locations to be interchangeable so experiment targets can be moved around the core to achieve variable flux levels or to access different neutron spectrum locations. These irradiation locations should have a larger experimental cross sectional area than available in HFIR (>1.5 inches diameter).

At least one closed cooling loop experiment location should be included to allow experiments to be isolated from the reactor cooling system. This design would allow the possibility of testing fuels to failure.

A fast spectrum experimental location is needed for materials irradiation. This site would likely use high-energy flux enhancement using fission plates or thermal neutron absorbers and would require a closed loop with a non-moderating coolant. This and other open-loop materials irradiation facilities should be instrumented with the capability of controlling capsule internal temperatures to greater than 350°C.

Other Options Considered

A fast reactor was considered but the consensus was that such a facility would be complicated and expensive to operate (e.g., FFTF). Most isotope production is better done in a lower energy neutron spectrum.

Neutron beams should not be a primary focus of the reactor but could be added if they do not interfere with the other principal functions of isotope production and materials irradiation.

Co-Location Options

Consider co-locating an accelerator (large cyclotron) to allow for production of isotopes that are more efficiently produced by these machines Build multiple reactors on the same site, one fast and one thermal spectrum or one isotope production machine and a beam reactor in order to take advantage of the infrastructure, or two thermal spectrum reactors operating on alternating schedules to ensure a steady, reliable supply of strategic isotopes Add a Critical Assembly similar to the Pool Critical Assembly (PCA) for training and use by students

Integrate a system of hot cells with the reactor pools to allow refinement of isotope materials and examination of irradiated materials within the same complex.

> **Design Options**
>
> The team recommended that the design begin with the ORR reactor configuration converted to use a high-density LEU fuel. Calculations will be needed to determine the flux levels that might be achieved with the addition of a closed loop and a fast flux capability.
>
> It was also suggested that a movable deuterium tank, which could be moved to one edge of the reactor, might provide additional flexibility with respect to flux level and spectrum. Graphite or Beryllium may also be options for such a reflector. As the design progresses, it will be necessary to prioritize the experimental characteristics in order to provide an optimum design.
> All of these characteristics are important when considering the design of the future reactors and experimental facilities. As isotope research shows more and more promise in so many different areas, the task of designing the next generation of reactor becomes more pressing. Time is running out to bridge the gap between the old and the new.

8. ISOTOPE HARVESTING FROM LONG-LIVED STOCKPILES

Potentially important sources of isotopes for research, development, and industry are found in existing isotope stockpiles and irradiated targets which were produced by previous DOE programs and/or were byproducts of the programs. Most of these programs have ceased functioning or the DOE industrial complexes no longer have the capacity to produce isotopes. In addition, many of the feedstocks no longer exist. To produce these isotopes again in the amounts previously produced and in the current regulatory environment could require a multi-billion dollar investment and a very long lead-time. These existing isotopes are unique and invaluable and will most likely never be produced again in these quantities. A number of years ago, DOE- EM conducted an inventory of isotopes present in the DOE complex. Some of the isotope inventory has been disposed of or is scheduled for disposal. However, a significant amount of very important isotopes remains in inventory. Some of these are classified, and therefore, cannot be discussed here. Listed below are some of the stockpiles and/or irradiated targets and the isotopes of particular interest.

^{233}U

Production Site: Savannah River and Hanford production Reactors

Inventory Site: Oak Ridge National Laboratory and Idaho National Engineering Laboratory. The ^{233}U material at ORNL has been separated from fission products and contains only decay daughters and added chemical elements. The material at Idaho National Engineering Laboratory consists of two UO_2-ThO_2 cores for the Shipping Port LWBR test. One core is unirradiated and the other was irradiated.

Production Mode: ^{233}U is most efficiently produced via neutron capture of ^{232}Th. The contamination of ^{233}U by ^{232}U is dependent on neutron spectra and the length of irradiation. The $_{232}$U decay daughter, ^{208}Tl, contributes the predominant penetrating radiation. ^{233}U can also be produced at a much smaller scale as a decay daughter of ^{237}Np and ^{233}Pa.

Isotopes: ^{233}U has a variety of applications and research interests. The most visible and important application of ^{233}U is as a "cow" for the production and recovery of ^{229}Th. The ^{229}Th is then processed for its ^{225}Ac decay daughter. The ^{225}Ac decay daughter can also be fabricated as a generator for $^{213}Po/^{213}Bi$. As discussed in Chapter 3.A and Sidebar 4.1, these isotopes are being studied for the treatment of myeloid leukemia. Ultra-high purity ^{233}U is also used for detection of U in environmental samples and forensic studies in the detection of diversion activities and covert production.

The DOE ORNL stocks have a Congressional mandate and are under contract for downblending for disposal beginning in 2012. If ^{229}Th extraction were permitted, it is estimated that about $4M would be needed for a proof of principle demonstration of the process, in large part to prepare the safety analysis and documentation. The subcommittee has included this demonstration in its optimum budget scenario as a technically proven path to begin addressing the need for ^{225}Ac and its decay products. Processing ever more ^{233}U material could bring the total cost to $20M. At the same time, the subcommittee recommends R&D in alternative production techniques, such as neutron irradiation or high-energy accelerator production from ^{232}Th.

^{241}Am in Excess of NNSA Stockpiles

Production Site: Los Alamos National Laboratory (Primary)
Inventory Site: Los Alamos National Laboratory (Primary)
Production Mode: ^{241}Am is produced through the beta decay of ^{241}Pu ($t_{1/2} = 14.29$ yrs). ^{241}Pu is produced in all reactor irradiations of U and Pu. Generally, though, the ^{241}Am is diluted by the presence of ^{243}Am. ^{241}Am can be produced essentially isotopically pure from the decay of separated Pu. The DOE complex has a significant inventory of excess Pu and NNSA Pu is a potential source of ^{241}Am. In addition, during a turn-down in the oil and gas industry, companies were allowed to return their ^{241}Am sources (typically 3, 5, or 20 Ci sealed sources).

Isotopes: Recovery of the ^{241}Am would require separation from significant quantities of Pu. Currently, there is no large inventory of separated ^{241}Am remaining in the United States. The last batch of separated ^{241}Am was sold to foreign customers. France, which has an active processing and separation complex, does not recover ^{241}Am. ^{241}Am has a multitude of uses. It is used extensively in well logging by the oil and gas industry. It is widely used in smoke detectors (the basis for the most recent sales). It also serves as source material for minor actinide transmutation studies and experiments, as the source of alpha particles for AmBe neutron sources (both fixed and switchable) and as target material for the production of ultra-pure ^{238}Pu (no ^{236}Pu contamination). At present there is only one foreign supplier. LANL has proposed reestablishing their ^{241}Am capability at the cost of about $7M to complete installation of their Chorine Line Extraction and Recovery process (CLEAR) and to produce the first 250 gm of product and $2-3M per year in operating costs thereafter. They are seeking industrial partners.

MK-18A Targets

Production Site: Savannah River Site
Inventory Site: Savannah River Site
Production Mode: ^{242}Pu targets irradiated in K-Reactor for ~10 years. The primary mission was to produce multi-gram quantities of ^{252}Cf. During the 10 year irradiation the targets were exposed, for one year, to thermal fluxes in excess of 5×10^{15} n/cm^2- sec. This flux is about twice what is produced at the ORNL HFIR. It is believed to be the highest steady state thermal neutron flux ever produced in a reactor. Sixty-five of the MK- 1 8A targets remain in storage at the Savannah River Site.

Isotopes: ^{244}**Pu:** ^{244}Pu is the longest-lived Pu isotope. Its production is virtually non-existent in all other reactors because of the extremely short half-live of its immediate transmutation precursor, ^{243}Pu ($t_{1/2}$ = 4.956 hrs). Due to this feature, ^{244}Pu is not present in reactor-produced Pu or weapons-grade Pu. This lack of ^{244}Pu in all other existing Pu makes it the perfect radio-tracer. It is critical to the fields of safeguards applications, detection of Pu in environmental samples, forensic studies in the detection of diversion activities and covert production, and processing and in the accurate measurement of declared reactor fluxes. Its very long half-life ($t_{1/2}$ = 8.0×10^7 yrs) and heavy mass make it a valuable target material for the production of super-heavy elements. Its long half-life also allows for 1) bench-top or "cold" experiments on Pu (Synchrotron irradiations, electron microscopy, training students with ^{244}Pu, etc.), 2) minimization of radiolytic effects, allowing for a better understanding of the fundamental chemistry of Pu, 3) the study of Pu in biology R&D (minimization of α and x-ray emission/damage), 4) the study of aging issues related to stockpile stewardship, 5) the production of ^{240}U from the decay of ^{244}Pu for use in calibration of detectors in homeland defense applications, 6) double beta decay experiments, and 7) the production of ^{247}Pu. The world's supply (~3 grams) of separated and enriched ^{244}Pu came from the processing of 21 of these targets and enrichment at ORNL. Another ~20 grams of unprocessed and un-enriched ^{244}Pu remain in the irradiated MK-1 8A targets. This is the world's inventory of ^{244}Pu. It will not be produced again in these quantities.

Heavy Cm: In the transmutation route for the production of ^{252}Cf, the Cm isotopes, ^{244}Cm through ^{248}Cm, are produced. Due to the long irradiation time and long out-of-reactor decay of the MK-1 8A targets, the Cm isotopic distribution has become heavier. In other words, a good deal of the ^{244}Cm has either been transmuted or decayed ($t_{1/2}$ = 18.1 yrs) and the ^{245}Cm has fissioned or transmuted to ^{246}Cm and heavier Cm isotopes. This "heavy" Cm isotopic distribution makes the Cm excellent target material for the production of the transcurium elements up to Fm. The yields of these elements are greatly enhanced relative to lighter Cm distributions because of the burnout of ^{245}Cm (σ_f = 2141 barns). In addition to the production of transcurium elements, the heavy Cm is a potential feedstock for the actinide enrichment of ^{246}Cm, ^{247}Cm and ^{248}Cm. Enriched ^{247}Cm would be ideal for radiochemistry and solid state actinide chemistry because of its long half-life ($t_{1/2}$ = 1.56×10^7 yrs).

^{250}**Cm:** The MK-18A targets may also be a source of ^{250}Cm ($t_{1/2} \sim 8.3 \times 10^3$ yrs). Its transmutation precursor is ^{249}Cm which has a very short half-life ($t_{1/2}$ = 64.15 mins). However, the very high thermal neutron flux and long irradiation times may have produced ^{250}Cm in quantities of interest. The principal use of this neutron-rich isotope would be as a target for superheavy element production.

MK-42 Targets

Production Site: Savannah River Site

Inventory Site: Oak Ridge National Laboratory

Production Mode: ^{239}Pu targets irradiated in the C-Reactor for ~4 years. The design irradiation goal was 87 atom % fission. The MK-42 targets are among the highest burnup targets in the DOE complex. Most of these have been processed for use by existing DOE programs but several unprocessed targets remain in inventory, and a number of processed Am-Cm fractions also remain in inventory. All of the remaining MK-42 inventory is at ORNL.

Isotopes: ^{243}**Am:** The ^{243}Am in both inventories represents essentially the entire inventory of ^{243}Am in the United States with the exception of several grams scattered throughout the DOE complex. The Am is ~70 atom % ^{243}Am. Inventory amounts are on the order of hundreds of grams. Potential uses of the ^{243}Am are as feedstock to future actinide enrichment devices, source material for minor actinide transmutation studies and experiments and as target material for the production of transamericium elements. It is one neutron capture removed from ^{244}Cm, and the yield of transcurium elements is roughly 10-15% less than Cm rich in ^{244}Cm.

Light Cm: Light Cm, Cm rich in ^{244}Cm, was produced in multi-gram amounts with inventories on the order of a couple hundred of grams. This inventory, along with some other SRS-produced Cm, essentially represents the total inventory of light Cm in the United States. The light Cm is suitable target material for the production of transcurium elements (the yield is lower than that for heavy Cm). Several years of irradiation in the ORNL HFIR could transmute it to heavy Cm which is an optimum target material for heavy element production. In addition to the production of transcurium elements, the light Cm is a potential feedstock for the actinide enrichment of ^{244}Cm and ^{245}Cm and source material for specific RTGs.

^{79}**Se,** ^{93}**Zr,** ^{99}**Tc,** ^{107}**Pd,** ^{126}**Sn and** ^{129}**I:** The MK-42 targets were designed for 87 atom % fission. Each target contains about 2.6-to-2.7 kg of fission products. In this fission product inventory there are multi-milligram to multi-gram quantities of very long-lived fission products of interest. Some of these are ^{79}Se, ^{93}Zr, ^{99}Tc, ^{107}Pd, ^{126}Sn, and ^{129}I . The isotopic distributions of these isotopes (produced by fission of ^{239}Pu) are far more attractive than what can be produced by direct transmutation of the lighter stable isotope precursor. The production and destruction of these isotopes in a nuclear reactor (neutron cross section studies) and their subsequent impact on waste disposal are of interest. ^{93}Zr has been recovered in multi-gram quantities during a campaign at the ORNL Radiochemical Engineering Development Center.

^{3}He

Production Site: Savannah River and Hanford production Reactors

Inventory Site: Savannah River Site and Savannah River National Laboratory, NNSA. Tritium supplies come from dismantlement and maintenance of nuclear weapons. ^{3}He is recovered following beta decay of tritium. The IPDRA program is responsible for the sale and distribution of the ^{3}He on behalf of the NNSA, based on a full-cost recovery basis. If the size of the stockpile is decreased in the future, the amount of ^{3}He available will also decrease.

Prior to 2001, the amount of ^{3}He recovered exceeded demand. Since 2001, demand has greatly exceeded supply and the excess supply of ^{3}He is almost depleted. SRNL has a memorandum of understanding with the isotope program to provide 10,000 l/yr for the next five years. As discussed in Chapter 3.B and 3.C, a significant shortage of ^{3}He is projected. The subcommittee does not see a ready solution to the ^{3}He supply issue, unless other neutron detector technologies are substituted for ^{3}He counters. One possibility that merits investigation is obtaining ^{3}He from Canadian CANDU reactors. An interagency working group (DOE/NNSA, DHS, DOE-ONP) led by DOE/NNSA is currently working to address this issue and better coordinate and inform each agency of the work being performed and the ultimate distribution.

Summary

These stockpiles of isotopes represent a precious resource for the nation and result from major national investments. The subcommittee recognizes the potential major environmental concerns and costs associated with the continued storage and maintenance of these stockpiles, and the need in many cases for long-term solutions that would make isotope recovery impractical. The subcommittee does urge that the unique nature of these isotopes be weighed heavily. In particular, the great potential for alpha-therapy brings the ^{233}U situation to the fore. In the first report the subcommittee presented ^{233}U as a possible interim solution that needs to be seriously considered for the short term until other production capacity can become available.

There remains industrial need for ^{241}Am. The realization of a commercial partnership to satisfy the commercial demand seems to be the appropriate course of action for this isotope.

9. RESEARCH AND DEVELOPMENT FOR PRODUCTION AND USE

At first glance the amount of research and development that is required for a "production program" might seem to be limited. However, as the previous chapters have illustrated, there are many examples where considerable R&D needs to be done, either to produce new isotopes efficiently, to capitalize on new opportunities, to take more complete advantage of existing production capacity, or to obtain the information to make realistic plans to deal with perceived shortfalls beyond existing technology. In order for the program to continue to evolve as a responsible and creative resource, there is a need to create an environment and capacity for research in areas that will meet the anticipated growing demand for isotopes in many basic and applied fields of study.

Some of this R&D can be performed in a short time scale with modest resources. Other areas may likely require a concerted effort over a 5-10 year time scale to provide significant results.

Radionuclide Research

Accelerator Production

In order to enhance capacity for producing radionuclides produced on accelerators there are four areas of study that can provide an increase in the availability:

- Increasing beam current of existing production routes.
- Identifying alternative routes to utilize existing production capacities using low-energy particles in new ways.
- Improving chemical processing for isolating desired isotopes.
- Developing new approaches for production isotopes.

The production of radionuclides is governed by the following equation:

$$Y = I \, n \, \sigma \, (E)*SF/\lambda$$

The yield (Y) is the number of nuclei produced, I is the intensity of the beam in particles/second, n is the number of target nuclei, σ (E) is the probability of production expressed as a reciprocal area (cm^2) and is a function of energy of the bombarding particle, and SF is the saturation factor ($1-e^{-\lambda t}$) which takes into account the fact that the radionuclide produced is radioactive and is decaying with a rate constant of λ ($= \ln 2/t_{1/2}$, with $t_{1/2}$ the half-life of the isotope of interest). t in the saturation factor is the length of irradiation. From the above relationships it can be seen that as the length of irradiation increases, there will be a point where the rate of production equals the rate of decay and no further production is possible; the production has reached saturation.

An increase in yield can be affected by any of the parameters or combination thereof. However, each parameter has its limit. The intensity of beam particles may be limited by capabilities of ion sources, activation of accelerator components by beam losses, or the ability of the target to absorb the heat deposited by the beam. In fact, heat dissipation is probably the single most important factor in limiting the production of radionuclides in accelerators.

In addition to exploring means for increasing beam current through providing better target materials, there is the possibility of exploring alternative production routes using low-energy particles. Having alternative routes via the use of deuterons or helium particles (^{3}He and .) may provide unique possibilities for increasing purity of the desired product. As discussed in Chapter 6, there may be excess capacity in low-energy PET isotope producing accelerators that could be brought to bear. Funded by a NIH/NCI grant, Washington University in St. Louis has demonstrated the ability to produce a number of research isotopes (Sidebar 9.1). What is required is more R&D to optimize and standardize targets and chemistry procedures.

Chemistry for Isolating Desired Isotopes

While it may seem obvious, chemistry is at the very heart of any isotope production process. Developing more efficient processes can also impact the environment by reducing toxic chemical and radioactive species released to the environment. As an example, Russian chemists have a long history in using thermochromatographic techniques for isolating elemental species [N090]. Typically, this approach makes use of the difference in vapor pressures of species as a function of temperature to separate and isolate the desired species without resorting to *wet* chemistry (using acids/bases and/or organic solvents) to extract the desired species.

Microfluidics may be able to play a role in examining alternatives. Using such devices the chemistry of different processes can be examined in a rapid sequence, saving resources and time. In fact, microfluidics might be used in the actual processing since the number of atoms of the desired species is typically in the subnanomolar range.

New Approaches for Production Isotopes

Accelerator technology continues to make significant advances and these advances call for a reexamination of whether more effective technologies are now available that would change the standard production paradigms. High current electron accelerators are being reconsidered for isotope production (for example [TR08]). Photons are the inverse of neutrons, that is the typical neutron reaction includes (n,), (n,p) or (n,), etc. while the photon interactions are (,n), (,p), etc. However, their interaction with matter is about 1000 times smaller than neutrons.

Technology of ion linacs and developments in consideration of accelerator driven fission systems have advanced to the point where they should be re-explored as alternatives to building new reactors for isotope production, either for fission product isotopes or other neutron-rich radionuclides.

While a community may wish to have a particular radionuclide for whatever purpose, it is incumbent for both researchers and the research funding agencies to ask the question: Can enough be made if the demand reaches desired goals? If a particular radionuclide shows such promise to be used clinically either in diagnostic or therapeutic situations, many curies per week will be required to sustain its usage (Sidebar 4.1).

Sidebar 9.1: New Uses for Low-Energy Cyclotrons

Washington University at St. Louis (WU) has an extensive program in the production of non-standard radioisotopes for Positron Emission Tomography (PET) using cyclotrons of maximum energy less than 20 MeV. Over the last 10 years Washington University has supplied over 60 institutions with exotic positron emitting isotopes (^{60,61,64}Cu ^{86}Y, ^{76,77}Br, ^{89}Zr, ^{94m}Tc) for academic research. This radionuclide resource has been critical for extensive medical research representing millions of dollars in DOE and NIH grants. The radioisotope production has resulted in almost 200 publications from research conducted with non- standard isotopes from Washington University.

The development of ^{64}Cu as an imaging radionuclide can be particularly noted as a success. The ^{64}Cu produced at WU continues to be used in a variety of projects both internally and in collaborative investigations. A clinical trial is about to begin where the ^{64}Cu will be incorporated into the compound ATSM and used to measure the hypoxic nature of tumors. Production of the radionuclides is on an ongoing basis, with ^{64}Cu being produced almost weekly and ^{86}Y, ^{89}Zr, ^{94m}Tc and ^{76}Br slightly less frequently.

WU has been shipping several of these radionuclides to other institutions that do not have cyclotron access for the past 10 years. The figure below shows the amount of these radionuclides shipped since 1999. This program is an outstanding example of how the innovative development of target and chemistry techniques could extend the isotope supply with a network of low-energy facilities.

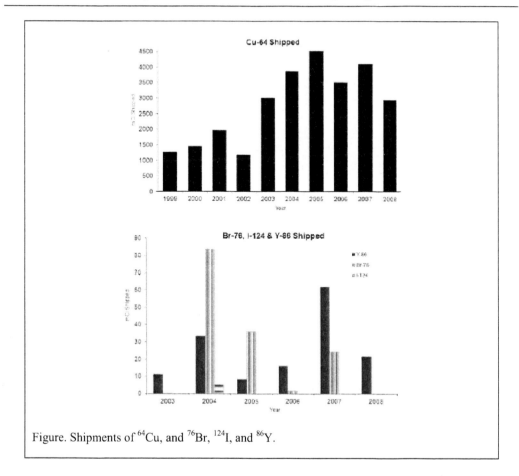

Figure. Shipments of ^{64}Cu, and ^{76}Br, ^{124}I, and ^{86}Y.

As part of this approach, there should be an assessment to determine whether alternatives exist, either in the form of different radionuclides that possess the appropriate physical and chemical characteristics to serve as a substitute or in the form of alternative methodologies that can be used.

Cross section measurements may have to be performed to determine the optimal conditions for production of the desired radionuclide, especially in the case of new alternative production approaches.

Most Compelling Opportunities and Impacts

The coordination of DOE accelerators along with the large accelerators at major laboratories around the world has proven to be most successful. To improve upon this approach the coordination and utilization of beam availability at both university PET cyclotrons and DOE accelerator facilities should be explored to enhance and expand the research isotope portfolios from these two types of facilities (low-energy and high-energy charged particle reactions).

In order to take advantage of such an approach, the development of target systems that can be utilized across different cyclotron platforms should be explored so that a unified target system can be provided to the target community for ease of use and eventual licensing. This is primarily an engineering task, but one should not undervalue the effort and cooperation required to achieve uniformity and consistency.

Finally the non-traditional accelerator approaches to isotope production (Sidebar 9.2 e.g., electron accelerators and increased use of high-energy spallation) will require development of new targets and subsequent chemistries.

What Are the Gaps and Hindrances for Reactor Produced Radionuclides?

The neutrons supplied from a reactor are an extraordinarily efficient resource for producing radionuclides, and R&D needs to be directed to expand this capability, either within the DOE envelope of resources or to enable isotope production at university research reactors.

As therapy using radiotoxic nuclides matures from a purely research perspective to a clinical reality, there will be growing demands on the availability of radionuclides that decay by ,,-minus emission. In order to meet this demand for high-purity radionuclides, a program for research and development in radiochemistry is needed to produce , and ,,-emitters with high specific activity in chemical forms targeting radioimaging and radiotherapy.

At the same time, as discussed in Sidebar 7.3, given the aging nature of the current reactor fleet, the nation needs to begin planning now for the isotope production reactors of the future.

Stable Isotopes

In order to maintain and increase availability of a wide array of stable isotopes that are used directly in research and as starting materials for other areas for research, there needs to be a reinvestment in developing and prompting new techniques and approaches to enriching isotopes. The advances in technology for ion sources and magnetic separation should be explored to reinvent the Calutrons.

Sidebar 9.2: Alternative Accelerator Approaches to ^{99}Mo Production

As mentioned elsewhere in this report, the commercial production of ^{99}Mo is not a subject addressed by the NSACI committee, but it is a major issue and the current fragility in the supply of this isotope threatens future clinical nuclear medicine practice. Clearly something needs to be done globally to strengthen the supply reliability of ^{99}Mo, and the DOE will have a role to play with respect to solutions for the United States. While the desired approach may be a low-enriched-uranium (LEU)-fueled reactor using LEU targets, there are not many (or any) candidate reactors that are not already being used for this purpose. The University of Missouri Research Reactor (MURR) has submitted a letter of intent to the Nuclear Regulatory Commission for licensing an LEU ^{99}Mo production facility and is doing conceptual studies. This reactor would remain HEU-fueled until advanced LEU replacement fuels are available.

The NSACI subcommittee is also aware of efforts by Babcock & Wilcox and Covidien for the development of a Aqueous Homogenous Reactor (AHR) for the production of ^{99}Mo. The AHR design would utilize LEU fuel for the production of ^{99}Mo, and the several units planned would be capable of supplying roughly 1/2 of the U.S. need for ^{99}Mo. There would be no need for any HEU in this operation.

Since both the short and long term prospects for reactor production of ^{99}Mo without the construction of a new reactor facility are uncertain, and the nation should also prepare for a potential significant increase in the ^{99}Mo demand, the subcommittee believes that R&D into alternate production approaches for ^{99}Mo is an area of significant interest and opportunity. Potential alternatives that would benefit from R&D include the following:

- The development of small, compact solution reactors where the uranium-containing solutions are both the reactor fuel and the isotope production target.
- Photo-fission of ^{238}U using intense photons generated by an electron linac, ^{238}U$(\gamma,F)^{99}$Mo.
- Photo-neutron reaction using intense photons generated by an electron linac, ^{100}Mo$(\gamma,n)^{99}$Mo.
- Create a neutron flux to mimic the reactor flux and geometry and LEU target irradiations at a proton spallation source.
- Direct accelerator production of ^{99m}Tc using the reaction ^{100}Mo$(p,2n)^{99m}$Tc.

While each of these alternatives has been the subject of some R&D, additional R&D could lead to optimization of these alternatives. One or more could lead to significant domestic supplies of ^{99}Mo, especially if supply reliability is valued over market price.

It would be very valuable to continue the development of high throughput separators such as the plasma separators which are promising candidates to handle bulk enrichment for high volume applications. If necessary, these could be followed by high enrichment using electromagnetic separation for those isotopes that require high isotopic purity.

The availability of highly enriched isotopes for target materials can assist in the development of radionuclides produced by either accelerators or reactors. Such approaches expand the possibilities for supplying these valuable materials. As discussed in Chapter 5, enriched stable isotopes for isotope production represent one of the primary components of the sales of the DOE isotope pool.

With the issue of ^{3}He availability it is important that alternatives to ^{3}He be sought for the high volume uses while at the same time processes for improved collection or alternative production of ^{3}He be developed.

Conclusions

This discussion represents generalities with few specifics. In order to focus on any aspect, one has to take into consideration the specific radionuclides or isotopes, so that the above topics can be brought to bear. The priorities for R&D include the following:

- New generation of isotope enrichment capability including continued development of plasma separators.
- New production techniques for ^{3}He.
- Target technologies and standard chemistry procedures to enable a network of accelerators and reactors from the university and private sector that can be used to produce research radionuclides.

- Targetry research such as improving heat dissipation and improved isolation of the desired product.
- Challenging the request for radionuclides with limited production capability with alternative species and methodologies. New paths to accelerator or reactor production of emitters for therapy are very important examples. This includes providing needed data for cross sections, material properties, and chemistry yields.

While some of this R&D should take place at the existing isotope production complex, a substantial part of this research program should be open competitively to the broader research community and to industry. Such an open program is essential to bring in new ideas and new people to the isotope production enterprise. Indeed, it should be noted that R&D is an important path to expanding the skilled isotope production workforce and retaining the most creative individuals in the program.

Recommendations

The considerations of this chapter lead, in part, to the following recommendations of the subcommittee:

Support a sustained research program in the base budget to enhance the capabilities of the isotope program in the production and supply of isotopes generated from reactors, accelerators, and separators.

Invest in workforce development in a multipronged approach, reaching out to students, post-doctoral fellows, and faculty through professional training, curriculum development, and meeting/workshop participation.

In Chapter 12, the budget implications of these recommendations will be explored.

10. TRAINED WORKFORCE AND EDUCATION

The previous chapters have dealt with the current isotope programs and new technical capabilities required for the future. Perhaps no factor is more important to a future successful and thriving program than the *availability of a skilled workforce,* educated in the underlying disciplines that make up the broad range of compelling research opportunities afforded by isotopes from biology, medicine, pharmaceuticals, physical sciences, homeland security, and industry. New advances in all the disciplines discussed are strongly dependent on the education of a skilled workforce that will make the new discoveries and expand the use of isotopes. Yet, there is a common overwhelming concern amongst all the professional societies and some of the government agencies that the numbers of students getting degrees in the areas that could contribute towards new advances are diminishing. Professional societies of nuclear medicine, medical physicists, nuclear chemists, nuclear physicists, and, for example, the American Physical Society (APS) and the American Association for the Advancement of Science (AAAS) have studied with alarm the decrease in the number of Ph.D.s awarded in their respective areas. Historically, a significant percentage of the expertise brought to the production of isotopes came from advances in nuclear physics and nuclear chemistry (Figure

10.1). As new developments were made in construction of better accelerators, more advanced instrumentation for radiation detection systems, and a more comprehensive use of data bases, they were put to use to advance applications of nuclear science from medicine, biology industry, and homeland security. Simultaneously, expertise from nuclear chemistry was brought to advance nuclear medicine by developing ways in which the isotopes could be introduced into humans and animals.

The new advances and developments in nuclear medicine initially came from individuals from traditional education backgrounds in engineering, nuclear chemistry and nuclear physics who helped develop today's programs in medical physics and radiopharmacy, and helped create a completely new broad class of experts involved in modern medicine. Procedures involving nuclear medicine applications today number in the 20 millions with some average annualized increase in the number of procedures of approximately 5% per year. Over 50% of the procedures involving radioactive isotopes are to assess blood flow or other cardiovascular studies. Other uses for nuclear medicine procedures include tumor mapping, visualization, and therapy (Sidebar 10.1).

There is a clear sense that as the needs for nuclear medicine procedures grow the number of trained experts is not keeping up. There are presently 56 ACGME accredited nuclear medicine residency training programs in the USA. ACGME (Accreditation Council for Graduate Medical Education) is responsible for the accreditation of post-MD medical training programs within the United States. There are approximately 4000 board-certified nuclear medicine physicians, and about 5000 radiologists who practice nuclear medicine on a daily basis. There are 20,000 nuclear medicine technologists, about two thirds of these practice in hospitals and the balance work in private offices in the U.S. The number of trainees from all the medical physics programs in the country falls short of the required number of medical physicists. This situation will become more critical since after 2012, the American Board of Radiology will require certification in one of the four subspecialties: Therapy, Nuclear, Diagnostic, and Radiation Safely. Sidebar 10.1 lists a number of the largest medical programs with residencies in medical physics. Memorial Sloan Kettering and the University of Wisconsin are the only medical schools in the country with dedicated medical physics departments.

As discussed in earlier chapters, another growing area in the use of isotopes is homeland security, including nuclear forensics. Nuclear forensics is a developing interdisciplinary field working closely with homeland security, law enforcement, radiological protection dosimetry, traditional forensics, and intelligence work to provide the basis for attributing the materials to its originators (Sidebar 10.2). It includes the analysis of nuclear materials recovered from either the capture of unused materials or from the radioactive debris following an explosion in order to identify the sources of the materials and the industrial processes used to obtain them. In the case of a nuclear explosion, nuclear forensics provides the ability to reconstruct the key features of the exploding device. Correct attribution of materials is believed to be a deterrent to terrorism since nearly all nuclear materials are properties of governments. A 2008 report on "Nuclear Forensics: Role, State of the Art, Program Needs" [AP08] published by the Joint Working Group of the American Physical Society and the American Association for the Advancement of Science highlighted significant workforce concerns.

Figure 10.1. Graduate students receiving hands-on training in accelerator based techniques.

Nuclear forensics relies heavily on physical, isotopic, and chemical analysis of radioactive materials carried out at a number of DOE laboratories in the U.S. and some IAEA government and university laboratories around the world. *The report highlights the fact that trained specialists in nuclear forensics are few and over committed with a large proportion of them close to retirement age with no adequate workforce availability to replace them.* The problems of a declining pool of technically competent scientists and the need for new technology are behind the report's recommendations that actions be taken by the U.S. government and industry to accelerate the training of appropriate personnel. The suggestion is to create funding for research at universities in cooperation with relevant laboratories, to create graduate scholarships and fellowships, and to fund internships at laboratories. The suggestion is to increase the number of Ph.D. students by 3-4/yr initially and then to maintain the skilled personnel level after an initial period of growth for ten years.

**Sidebar 10.1: Workforce for Nuclear Medicine:
Growing Needs and Few Training Grounds**

Nuclear medicine is a highly multidisciplinary specialty that develops and uses instrumentation and radiopharmaceuticals to study physiological processes and to diagnose and treat diseases. Modern nuclear medicine benefits from the production and use of a multitude of stable and radioactive isotopes as well as from partnerships with national laboratories, universities, and industry in developing the facilities for isotope production, chemical techniques for synthesizing radiopharmaceuticals, and instruments that can detect the radiation for imaging. All aspects depend on a skilled workforce in the various subdisciplines, most especially nuclear medicine physicians and technicians, engineers, nuclear chemists, and physicists.

An estimated 19.7 million nuclear medicine procedures were performed during 17.2 million patient visits in the United States in 2005, in over 7,200 hospital and non-hospital sites (according to a report just released by the IMV Medical Information Division). These numbers represent a 15% increase for 2002 to 2005, for an average annualized rate of increase of 5% per year. Over half of the patient visits were for cardiovascular studies, including cardiac perfusion.

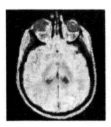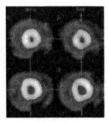

Blood flow with radiopharmaceuticals.

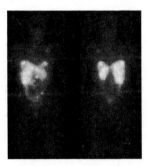

Tumor mapping and visualization by radioactive isotope accumulation.

Some of the largest medical schools that offer residencies in medical physics are:

Memorial Sloan Kettering	University of Wisconsin
University of Maryland in Anderson	Washington University
University of Chicago	University of Indiana
University of Michigan	University of Washington

Only Memorial Sloan Kettering and the University of Wisconsin have dedicated medical physics departments.

The report suggests drawing personnel from *geochemistry, nuclear physics, and nuclear engineering, materials science, and analytical chemistry*. The increase in a trained workforce is coupled with recommendations for investment in the development and manufacture of advanced laboratory equipment.

Sidebar 10.2: Workforce for Nuclear Forensics

Nuclear forensics is the science and technology of attribution. It is both invaluable intelligence for responsible decision making and a deterrent to acts of terrorism using nuclear materials. Forensic science involves the analysis of nuclear materials, whether intercepted intact or retrieved from post explosion debris, to identify the origin of the material and the processes it has undergone. Tools of nuclear forensics come from a variety of disciplines: nuclear engineering, analytical and nuclear chemistry, nuclear physics, geochemistry, and computation sciences. Major aspects of this work require the use of and the intimate knowledge of isotope tools. The figure below shows an example of a potential malicious act over the island of Manhattan. Response teams that would handle such emergencies would be trained in nuclear forensics.

The 2008 report from the American Association of the Advancement of Science, the American Physical Society, and the Center for Strategic and International Studies [AP09] identified two key initiatives to improve the capabilities of nuclear forensics: 1) the implementation and development of advanced equipment and 2) the availability of a skilled workforce.

A common theme in all the recent reports, including those specifically addressing the decrease in the number of Ph.D.s awarded to nuclear physicists and nuclear chemists, is the severe shortage of a skilled workforce to carry on the present status of research without even considering advances. The 2004 NSAC report on "Education in Nuclear Science" [NS04], the 2007 National Academies report on "Advancing Nuclear Medicine through Innovation" [NR07], the 2007 Long Range Plan for Nuclear Physics [NS07], the Nuclear Forensics report [AP08], the joint AAAS, APS and CSIS (center for strategic and international studies) report on "Nuclear Weapons in the 21st Century and U.S. National Security" [AA08], and the report of the APS panel on Public Affairs Committee on Energy & Environment "Readiness of the U.S. Nuclear Workforce for 21st Century Challenges" [AP08] all point to the shortfall in the availability of the required expertise to carry out jobs in the present situation with even more dire predictions for future growth. At a time when there is tremendous growth in the applications of nuclear science and growing societal challenges from energy to health and homeland security, there is a sharp decrease in the production of nuclear scientists from academia. Figure 10.2 below shows the age profile of nuclear and radiochemists (courtesy of Mark Stoyer and the ACS Division of Nuclear Chemistry and Technology). Figure 10.3 illustrates the decline in the number of PhD's awarded in nuclear science since the mid 1990's.

The NSAC white paper "A Vision for Nuclear Science Education and Outreach" [NS07A] written for the 2007 Nuclear Physics Long Range Plan recognized the potentially serious shortage the nation faces of trained workers in pure and applied research, nuclear medicine, nuclear energy, and national security. They cite from the "Education and Training of Isotope Experts" report published in 1998 and submitted to Congress by the AAAS that noted *"Too few isotope experts are being prepared for the functions of government, medicine, industry, technology and science."* Following a charge on education in 2004, NSAC recommended, following an extensive survey of the nuclear science workforce, that efforts be made towards a significant increase (20%) in new nuclear science Ph.D.'s for the next 5-10 years in order to meet the needs of the nation with a skilled workforce.

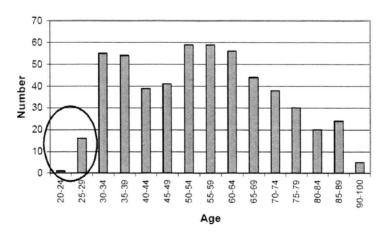

Figure 10.2. The age profile of members of the ACS Division of Nuclear Chemistry and Technology. The gap at the 25-29 and the 20-24 age distributions is a significant indicator of skilled workforce availability for the coming years.

The 2007 National Academies report on "Advancing Nuclear Medicine Through Innovation" recognized that nuclear medicine today is an interdisciplinary enterprise involving biologists, chemists, physicists, engineers, pharmacists, and clinician scientists. The report recognized that the development of new agents will require the collaboration of molecular, cellular, and structural biologists; bioinfomatics specialists; and synthetic and radiopharmaceutical chemists. Improvements in instrumentation for combined modalities of imaging for animals and humans require highly specialized and trained medical physicists, nuclear physicists, and engineers. The development and maintenance of new cyclotron-based research and clinical facilities will require additional radiochemists, radiopharmacists, and physicists. The National Academies report, similar to this subcommittee's efforts, gathered data from scientific societies, government agencies, and industry. In each case, the findings were of significant shortages of skilled personnel in chemistry, pharmacy, physics, computer science, and engineering. An overall shortage of nuclear medicine personnel was seen as an impediment to advancing nuclear medicine through innovation. The report noted a pressing need for additional training programs with the proper infrastructure.

The reasons for the decrease in the number of nuclear physicists and nuclear chemists have been studied and while the list below is not a comprehensive one, it addresses the most significant ones.

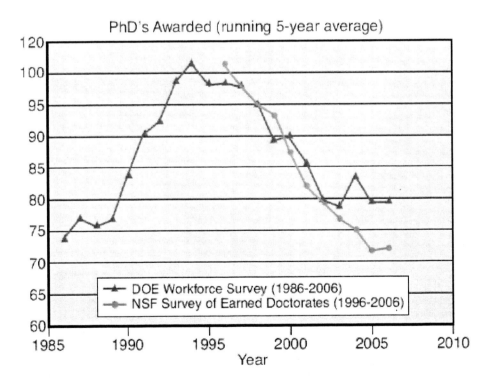

Figure 10.3. The number of Ph.D.s awarded 1986-2006 in nuclear science. Data are for students with full or partial support from DOE Nuclear Physics, and NSF Nuclear Physics programs summing those self-identifying as being in nuclear physics or chemistry. Running 5-yr averages have been shown to eliminate fluctuations.

A primary issue is a change in research priorities in nuclear physics and nuclear chemistry away from low-energy nuclear areas. As in all disciplines, as new research opportunities emerge, they attract a larger fraction of the resources. In the 1 970s essentially all of nuclear research (see for example the 1977 report of the National Research Council Committee on Nuclear Science [NR77]) involved the use of tools and techniques that could be applied to isotope applications. In the 21st century [NS07], this fraction has shrunk to roughly 20-30% as measured by federal funding. A major reversal in this trend is planned with the decision of the Department of Energy to build the Facility for Rare Isotope Beams at Michigan State University. This approximately $550M facility will give U.S. scientists world-leading capabilities in low-energy nuclear science, and indeed, one core of its research mission is to encourage development of isotope applications.

Universities offer a high exposure to potentially bright minds to pursue careers in these areas of national need. The broadening of vistas in chemistry and physics has led to a curriculum gap for undergraduates. There are very few if any courses offered to undergraduates in either nuclear chemistry or nuclear physics. Chemistry departments in academia have moved away from hiring in the area of nuclear chemistry. The same is true for low-energy nuclear physics. These are the sub-fields of these disciplines of the most relevance for applications of nuclear science. Many physics departments in the country do not have a nuclear physicist on the faculty, and physics majors do not have the opportunity to take a class in nuclear physics.

This void creates a disconnect. There is a growing need for nuclear chemists in nuclear medicine, radiopharmacy, energy, and nuclear forensics, yet academic positions for nuclear chemists have become nearly extinct. Despite the clear need for radiochemists, a graduate student entering a Ph.D. program in chemistry today is unlikely to be exposed to any formal training in radiochemistry (only 22 PhD programs out of 235 in the U.S. include nuclear chemistry at any level). Historically, if radiochemistry is included in a curriculum, it is part of inorganic chemistry, thus creating hurdles for interdisciplinary research, particularly for students who want to focus on areas outside of inorganic chemistry. Most chemical graduate education programs cannot adequately prepare scientists for modern, interdisciplinary research using radiochemistry.

The situation is similar in nuclear physics where there has been a significant reduction of university based accelerators from the approximately 30 operating in the early 1990's, resulting in less hands-on training of graduate students with expertise in related instrumentation development and accelerator advances. Today only one large and six medium or small nuclear physics accelerator laboratories are based at universities: Michigan State University, Triangle Universities Nuclear Laboratory, Texas A &M University, Yale University, University of Notre Dame, Center for Experimental Nuclear Physics and Astrophysics (U. Washington, Seattle), and Florida State University. A survey of these last six institutions prepared for the 2007 Long Range Plan meeting in Galveston (courtesy of M. Wiescher) shows the distributions of job choices by the Ph.D. students graduating from these 6 laboratories (81 Ph.D.s in the period from 2001-2006) in Figure 10.4.

Another major contributing factor has been the uncertain future of nuclear power in the U.S., leading to a significant decline in the training of nuclear engineers and the contraction of related nuclear technology.

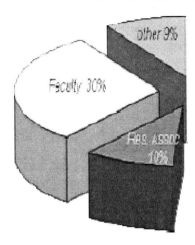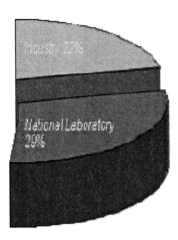

Figure 10.4. Distribution of career choices by Ph.D. students graduating from the six smaller university laboratories.

An issue specific to shortages in medical physics is the changing regulatory requirements.

The NSACI subcommittee considered various recommendations whose spirit is captured in our recommendation that the federal government assume significant responsibility for educating the next generation of nuclear scientists and engineers. The areas that must be addressed as a whole are many and include high school curriculum development, university level curriculum development, public outreach for the isotopes program to publicize the societal benefits and impact of the program, public broadcasting documentaries on isotopes such as NOVA programs, continued support of summer schools, distinguished named fellowships for graduate students, as well as postdoctoral fellows.

The discussion in this section illustrates that the issue of a shortage of a skilled workforce in isotope technologies is broadly based, but specifically impacts the future of the isotope program. Our recommendations in isotope production and development follow the broad recommendations of the NSAC Education report. However, the isotope program must focus its resources on the specific support of workforce development related to the program's mission, isotope production technologies.

Recommendation

Invest in workforce development in a multipronged approach, reaching out to students, postdoctoral fellows and faculty through professional training, curriculum development, and support for meeting and/or workshop attendances.

The subcommittee also recommends that industrial concerns that benefit from the isotope economy expand undergraduate internships, cooperative education opportunities, and training at their facilities in order to increase the pool of interested science and engineering students towards the development of a skilled workforce.

In summary, whether the topic is biology, medicine, basic physical sciences, homeland security, or energy, the shortfall of a skilled workforce in isotope development and production techniques is of paramount importance.

11. PROGRAM OPERATIONS

The national isotope program has long operated as a dispersed entity of the Department of Energy (and its predecessor agencies) with capabilities for isotope production and preparation at various facilities, primarily the national laboratories. This extensive capability evolved from the war-time Manhattan project and the development of facilities (Calutrons, reactors, accelerators) to enrich isotopes found in nature, initially uranium, and to produce new ones, initially plutonium. As discussed earlier, the end of the Manhattan project led to the use of this extensive isotope separation and enrichment capability to build up a storehouse of enriched stable and actinide isotopes for the research community, under the auspices of the Atomic Energy Commission. The development of major reactor and accelerator facilities provided a comprehensive suite of isotope production capabilities. As AEC evolved into other agencies, the capability to produce enriched stable and radioactive isotopes for national needs remained a strong legacy that led to new areas of research and development.

Much has changed in the operation of the isotope program in the past 25 years. Responsibility for the oversight and funding of the program has moved within the Department of Energy. With these changes in program location have come different rules and motifs for operation of the program. For decades the isotope program was operated as a resource for the research community, with researchers using materials on loan or paying only partially for the cost of production of the material. For many years there had been an effort to transfer the production of certain isotopes to commercial entities, dating back to the 1 960s. For example, Union Carbide, which managed Oak Ridge National Laboratory for over 30 years until 1984, built a small reactor to make and separate ^{99}Mo and later sold that process to another company. The move to Nuclear Energy within DOE in 1990 was accompanied by the dictate of full cost recovery and essentially no subsidies for the research community, along with increased emphasis on avoiding competition with private industry in the production of isotopes. Costs of most isotopes rose and DOE production of many isotopes ceased. Commercial suppliers often have imported enriched isotopes made in Calutron-like facilities, reactors, and accelerators in the former Soviet Union and elsewhere. This evolution has led to a large dependence on foreign sources for the production of many isotopes that are too expensive or unavailable in the DOE program.

At present the isotope program is funded by a mix of appropriated funds and sales of stable and radioactive isotopes. The uses of isotopes discussed in Chapter 3 are extensive and benefit many fields of research and development. As shown in Figure 11.1, medical research and applications are the single largest area of use of isotopes - 60%; 20% of sales are made to commercial entities, while the remaining 20% of isotope sales are made for an amazing array of research uses in many fields.

The mission of DOE's isotope program is threefold:

- Produce and sell radioactive and stable isotopes, associated byproducts, surplus materials, and related isotope services.
- Maintain the infrastructure required to supply isotope products and related services.
- Conduct R&D on new and improved isotope production and processing techniques.

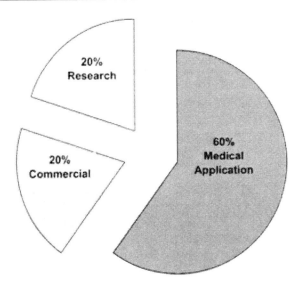

Figure 11.1. Distribution of recent sales of isotopes by the DOE isotope program in broad categories.

A glimpse of the breadth of isotope production and the some of the key uses of isotopes are shown in Figure 11.2. The nine sites shown produce and sell an amazing array of isotopes for a plethora of uses. The isotope program's 60% sales market for medical applications is clear when looking at the details of the sample inventory from these nine sites. Key radioisotopes are used in imaging and treatment of cancer and related diseases. Fundamental research in many areas depends on stable and radioactive isotopes. Lasers, fusion reactors, fundamental research, and detectors for national and homeland security applications need an increasing amount of ^{3}He.

In 2009, only three of these facilities (Brookhaven, Los Alamos, Oak Ridge) currently receive operating funding by federal appropriations to the isotope program. Others receive funding from isotope sales (Pacific Northwest, Idaho) or provide material that is sold by the program (Savannah River). Three represent the larger university and commercial sector and have historically coordinated with or have memoranda of understanding for potential collaboration (Trace Life Sciences/NuView in Denton, University of California-Davis, Missouri University Research Reactor). As discussed in Chapter 4, the operation of this broad program is complex and in many ways challenging, due to breadth of capability spread over many sites. This chapter examines the current status of the operation of the program and makes recommendations on future operational strategies.

The total sales revenue from this broad program was $13.6M in FY06, $15.3M in FY07, and $17.1M in FY08. A total of 88% of the revenue in FY08 came from sales to domestic entities (for research and commercial uses), an additional 6% resulted from sales to DOE or its contractors, and the remaining 6% came from sales to foreign entities. Sales of *stable* isotopes to foreign customers are a significant part of the business, representing 57% of the 171 stable isotope shipments in FY08. Conversely, 86% of the 391 shipments of radioactive isotopes went to domestic customers. Overall, only 10 customers provided almost 89% of all sales revenue.

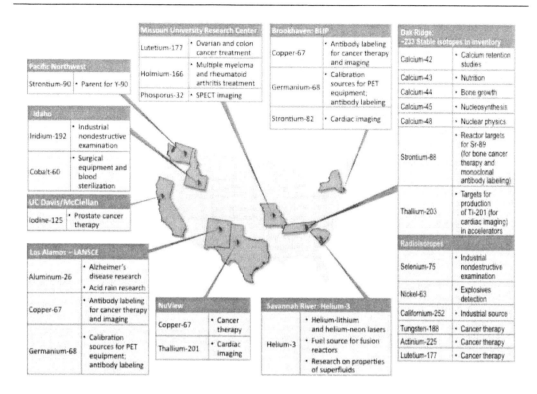

Figure 11.2. A partial landscape of isotope production including selected federal, university and commercial facilities. The isotope program has stewardship responsibilities for isotope production only at BNL, LANL, and ORNL. Some of the facilities depicted are non-federal facilities that may coordinate production with the isotope program.

The national laboratories have different and, in many cases, complementary isotope production capabilities. This is summarized in Figure 11.3 for the distribution of materials (by sales revenue) in FY08 from the various labs. The attribution to a single laboratory is somewhat notional. For example, in some cases targets irradiated at one laboratory are processed at another. These distributions can change significantly yearly. With these caveats, radioisotopes produced at Los Alamos accounted for 26% of the materials sold, at Brookhaven 24%, Oak Ridge 14%, and Pacific Northwest 2%. The electromagnetically separated stable isotopes stored and managed at Oak Ridge accounted for 7% of materials sold in FY08. The revenue for the isotope program in FY08 was, therefore, distributed over the national laboratories in the following way: 30% from material and services at Oak Ridge, 28% from Los Alamos, 25% from Brookhaven, 12% from Savannah River, 2% from Pacific Northwest, and 2% from BWXT Y-12 (included in Figure 11.3 in technical services and stable isotopes).

There are a few popular isotopes that dominate the market. For example, 38% of the FY08 revenue resulted from the sale of ^{82}Sr, produced at Brookhaven and at Los Alamos; 12% from ^{3}He made at Savannah River; 8.4% from ^{252}Cf made in HFIR at Oak Ridge; and 7.4% from ^{68}Ge made at Brookhaven and at Los Alamos. This distribution is shown in Figure 11.4.

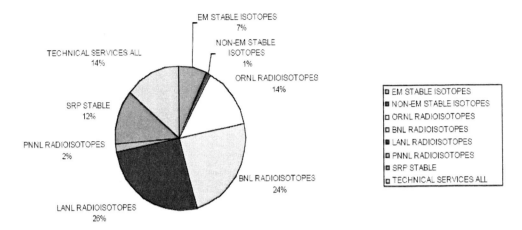

Figure 11.3. FY08 isotope material type from the various national laboratories. The percentages refer to fraction of the sales revenue.

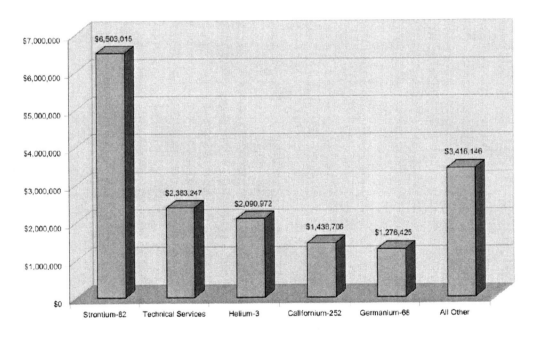

Figure 11.4. Revenue from the most popular isotopes in FY08.

As with any other production and sales organization, there are changes in the revenue and in the isotopes that generate the most sales. As shown in Figure 11.5, revenues from the isotope program have grown from $10M in FY04 to $17.1M in FY08. In 2004 the most popular of the big sellers was ^{3}He followed by ^{252}Cf and ^{82}Sr. By 2008 the sales of ^{82}Sr had surged to the top of the list, followed by ^{3}He. Also shown in this figure are the projected sales for FY09. Primarily because of ^{82}Sr and ^{68}Ge, revenue from accelerator produced isotopes is largest compared to revenue from reactor produced or stable isotopes. This trend has been increasing in the last three years, and in FY08 the numbers are 53% from accelerator-produced isotopes, 23% from reactor production, and 21% from stable isotopes.

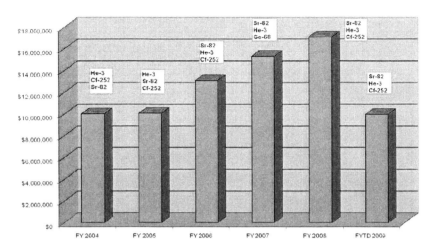

Figure 11.5. Sales in last five years in as-spent dollars, and the listing of the three top sales isotopes each year. The FY09 number is a projection made in April.

There were over 190 customers in FY08 resulting in more than 560 shipments. Although the largest percentage of sales revenue occurs for accelerator-produced radioisotopes and 'only' 21% of FY08 revenue was from stable isotopes, one can see in Figure 11.6 that stable isotopes have a higher percentage of shipment volume - 32%. Stable isotopes are generally bought in smaller volumes compared to reactor or accelerator-produced isotopes. The total number of shipments was 510 in FY06, 484 in FY07, and 562 in FY08.

The total annual resources combined from sales and DOE appropriation for the isotopes program ranged (in FY09$) from $29M in FY04 to $32M in FY08. The budget from isotope sales has grown from $11.9M in FY04 to $17.2 M in FY08 (in FY09$), which is an important trend, as shown in Table 11.1. In contrast, the appropriated budget declined over the same period (in inflation adjusted dollars), from $17.3M to $14.7M. This table also shows an important analysis of the portion of the budget that is re-invested in isotope research and development or in replenishment of the valuable infrastructure at the isotope production sites. An increasing amount of the sales revenue has recently been used to maintain the infrastructure, as high as $1 .7M in FY07 and $1 .3M in FY08 (in FY09$). This aspect of the program is very important, but unfortunately does not address the full set of needs for a generally aging infrastructure.

Reinvestment in R&D is also very important, e.g., to evolve new ways to produce isotopes of increasing importance to users. Generally this R&D expenditure has been small or zero, but a significant amount ($1.2M) was expended from the sales revenue in FY07. The investment of appropriated funds into R&D was apparently very small over this time period, not enough to register in this table. As the isotope program evolves in the future, it is important to strive for a regular and substantial investment into both infrastructure improvement and R&D.

The FY09 President's budget request for the isotope program increased in FY09 to $19.9M, primarily because of the addition of $3.1 M in isotope R&D and production funds. The Office of Nuclear Physics guided the subcommittee to use this level as the basis for projecting a constant effort budget into the future. The final FY09 appropriation was $24.9M. Table 11.2 shows how the request and the appropriated budgets are distributed across the three major national laboratory isotope programs at Los Alamos, Brookhaven, and Oak Ridge. Note that this appropriated funding provides support for a total of 32 FTEs across the national laboratories.

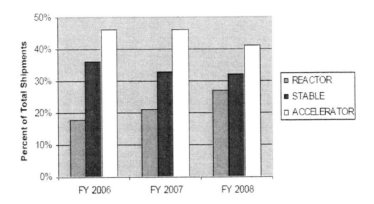

Figure 11.6. Percentage of shipment of isotopes by generation category in three fiscal years.

Of course, this is only part of the story, as the sales revenue also supports people working in the isotopes program. The distribution of FTEs supported by sales revenue and by the DOE appropriation is shown for the six national labs in Table 11.3. Note that people working on isotope production at Idaho, Pacific Northwest, and Savannah River are supported only from the revenue generated by those sales. At Brookhaven, most of the people are supported by the appropriated budget, while at Oak Ridge and Los Alamos there is a more even split between appropriation and sales revenue. This is complicated further by the fact that the FTE support is sometimes spread over many people, as services are purchased from the host national laboratory.

Table 11.1. Total funding of the isotope program over five fiscal years in inflation adjusted FY09 dollars. The total is split into an operating budget from sales and appropriations and investments in infrastructure and R&D

Thousands of Dollars	FY04	FY05	FY06	FY07	FY08
Total funding	$29,278	$27,108	$29,979	$32,381	$32,180
Operating budget from sales	$11,832	$11,111	$14,198	$13,152	$15,561
Operating budget from appropriation	$16,989	$15,191	$15,340	$15,999	$14,280
Sales invested in infrastructure	$115	$333	$441	$1,685	$1,260
Approp. invested in infrastructure	$342	$472	$0	$366	$665
R&D investments from sales	$0	$0	$0	$1,179	$414
Sales invested in infrastructure	1.0%	2.9%	3.0%	10.5%	7.3%
Approp. invested in infrastructure	2.0%	3.0%	0.0%	2.2%	4.5%
Sales invested in R&D	0.0%	0.0%	0.0%	6.9%	2.8%
Reinvestment - % of total budget	1.56%	2.97%	1.47%	9.98%	7.27%

Isotopes for the Nation's Future: A Long Range Plan

**Table 11.2. FY09 President's Budget request and appropriated
funding for the isotope program in FY09 and the FTE's funded
at each laboratory by the appropriation (rounded to nearest FTE)**

Lab	Total	Item	FY09 k$ Pres. Request	FY09 k$ Appropriated	FTEs
		Research Development and Production	3090		
		Research Development and Production - Production (estimate)		2430	
		Research Development and Production - Research (estimate)		2430	
		Other Research - SBIR/STTR	90	200	
		Associated Nuclear Support - including University Operations	750	870	
LANL	4640	IPF Operations/LANL Hot Cells	3650	3650	12
		IPF Upgrades	990	2490	
BNL	3470	BLIP Operations/BNL Hot Cells	3200	3200	8
		BNL Upgrades	270	270	
ORNL	7860	ORNL Hot Cells - Radioisotopes	3800	3800	3
		ORNL Chemical and Material Laboratories - Stable Isotopes	3764	3764	9
		ORNL Upgrades	296	1796	
Totals			19900	24900	32

This distribution of people supported by the two types of funds presents one of the challenges for the isotope program. The appropriated funding tends to be more stable than the sales revenue. While some of these sales-related FTEs should be purchased as needed for the variable sales achieved, maintaining mission readiness and far greater stability for the isotope program requires a somewhat larger number of FTEs to be supported on the appropriated budget. Each of the national laboratories has made this request, and this issue is discussed more fully below as the programs of each of the three major national labs are considered.

**Table 11.3. FTE support at the national laboratories
by the appropriation and by sales revenue**

	Total	Appropriation	Sales
ORNL	24.4	12.0	12.4
LANL	20.0	12.0	8.0
BNL	9.0	8.0	1.0
INL	0.2	0.0	0.2
PNNL	1.2	0.0	1.2
SRS	1.5	0.0	1.5

In 2009, the *American Recovery and Reinvestment Act* (ARRA) provided a most welcome new, one-time funding for two projects in support of the isotope program: Enhanced

Utilization of Isotope Facilities and R&D on Alternative Isotope Production Techniques. The first project will enhance isotope production and processing capabilities to better meet the needs of the nation for isotopes in short supply to industry and basic research. The second will improve America's competitiveness by investing in isotope production research at universities and laboratories. Both will contribute to training the scientific and technical workforce the U.S. needs. The final allocation of these funds, which are distributed in FY09 and must be spent by FY15, is still being finalized during the course of the writing of this report. The plans presented by ONP in March 2009 for investment in the isotope program are given in Table 11.4. These are particularly timely given some of the urgent infrastructure and research and development needs.

The subcommittee asked each laboratory to describe its current capabilities, its requests for resources for future operations, necessary and desired infrastructure improvements and new capabilities. The subcommittee does not endorse these individual requests, but includes them as examples of the needs and plans that must be considered to operate the program. The Office of Nuclear Physics will, as it does for all its facilities, conduct individual reviews of the isotope operations to plan for future operation and investment. In this year of transition from NE management to ONP, these reviews had not yet taken place.

Oak Ridge National Laboratory

This program includes stable isotope, reactor-produced isotopes, and the data center and business office. Radioisotopes are produced in HFIR and nearby hot cells are used to prepare isotopes for shipment. Over 200 stable isotopes are provided from ORNL as off-the-shelf products in various chemical forms.

Table 11.4. Preliminary allocations of the *American Recovery and Reinvestment Act* funding to the isotopes program

Purpose	Funding ($M)
BNL: Isotope Production Operations at BLIP	1.500
LANL: Isotope Production Operations at IPF	1.500
ORNL: Isotope Production at HFIR	0.725
INL: Isotope Production at ATR	0.700
BNL: Purchase and install inductively coupled plasma mass spectrometer (ICP-MS)	0.225
INL: Prepare cost estimates for isotope production	0.150
LANL: Replace 6 manipulators at 4 hot cells ($600K); refurbish all hot cell windows ($200K); upgrade hot cell electrical system ($500K)	1.300
ORNL: Purchase of stable isotopes to replenish inventory	1.000
ORNL: Remote Target Fabrication System refurbishment at REDC (^{252}Cf, ^{63}Ni, etc.)	0.900
ORNL: Replacement of the PaR Remote Handling System used for fabrication and container loading of ^{252}Cf (wire) sources	2.000
University and Laboratories: R&D for Alternative Isotope Production Techniques	4.750
TOTAL	$14.750

Custom chemical conversions and physical form preparations are available using metallurgical, ceramic, or vacuum processes to provide most stable isotopes in the desired forms for customer applications. Enriched stable isotopes are also often used as the precursor for the production of various radioisotopes.

The distribution of the 24.4 FTEs in FY09 across these various business elements and split between sales and appropriations is shown in Table 11.5. As noted above, 12 FTEs are funded by the appropriation, 12.4 by sales revenue. While the data center and associated business office are supported only by appropriations, activities associated with reactor operations at HFIR and use of hot cells for extracting radioisotopes are funded mostly by sales revenue.

The ORNL request for FY11 is to increase the total FTE count by 1.75, in addition to moving 2.4 FTEs from sales revenue to appropriated funds, and would lead to an increase of 4.15 FTEs at ORNL on appropriated funds. One problem they identified with the current mode of operation is the difficulty of planning the sales-related operation relating especially to use of HFIR and the associated hot cells when the volume of sales is difficult to project. Smoothing out the fluctuations in the sales revenue and planning a more stable operation would be greatly aided by the increase in FTEs funded by the appropriation and a decrease in those supported by sales. An increase in sales revenue should result from this mode of operation.

A consistent problem across the isotope program is the challenge of maintaining an infrastructure that is sometimes aging and certainly in need of replenishment in most cases. This is a problem at each of the national laboratories, and the ORNL needs are significant. The largest needs are discussed below, and, in some cases, the needs are being met already in 2009 by virtue of the *American Recovery and Reinvestment Act*. The list of immediate infrastructure needs presented by ORNL totals $1 1.9M, and $2.9M of this is being provided by ARRA funds:

Table 11.5. ORNL FTEs supported by the appropriation and by sales in FY09

FY09 FTEs by Unit - ORNL	Approp.	Customer	Total
Business Office	1.0	0.1	1.1
Nuclear Materials Processing	0.0	3.5	3.5
Isotope Development Group	4.0	2.3	6.3
Hot Cells	1.9	3.5	5.4
HFIR	-	3.0	3.0
Calutron Surveillance	0.8	-	0.8
National Isotope Data Center - Isotope Business Office	4.2	-	4.2
Total	12.0	12.4	24.4

- Target Fabrication Refurbishment - ensure operability of remote target fabrication needed for next transcurium campaign to start in early FY10.
- PaR Remote Handling - replace remote handling equipment used for Cf source fabrication and transloading.
- New Remote Target System - modernize and improve the fabrication of targets for HFIR including heavy element targets.

- Stable and radioisotope equipment replacement and upgrading, variety of equipment needed for processing, analyzing, fabricating, dispensing, and shipping radioactive and stable isotopes.
- Type B Cask - large shipment of radioactive materials: this is a DOE-wide issue and will be discussed below.

Maintaining current levels of production was presented to have the immediate needs that are listed above. The ongoing production of ^{252}Cf, ^{188}W, ^{63}Ni, ^{75}Se, ^{177}Lu, ^{225}Ac at current levels is expected to have a need for ongoing infrastructure investment on an annual basis to replace and repair old equipment, beyond the specific items listed above. These ongoing annual investments are estimated to be $2M per year.

In response to the subcommittee's request, ORNL presented options for expanding the production of several of these isotopes that would require significant infrastructure investment ranging from several million to tens of millions of dollars and the actual list of isotopes that are requested for expanded production. ORNL could produce more ^{225}Ac, ^{177}Lu, and ^{188}W and add new ones such as Bk, Es, and Fm from the Am/Cm targets now made.

For example, ^{225}Ac is being supplied by ORNL to support Phase II clinical trials of radioimmunotherapy for Acute Myeloid Leukemia (AML). AML is treated by injecting patients with approximately 10 mCi of ^{213}Bi which has been attached to an antibody that seeks the cancer cell in the body. A treatment consists of three injections in a day. The ^{213}Bi (46 min half life) comes from the decay chain of the ^{225}Ac (10 day half life). ORNL currently provides 100 mCi of ^{225}Ac every two months by milking a ^{229}Th cow at the REDC facility. This ^{225}Ac is used at the clinic to provide enough ^{213}Bi to treat up to five patients. Currently ORNL provides enough material to treat up to 30 patients per year. In addition to AML, many other types of cancer and HIV are being investigated for treatment with targeted ^{225}Ac/^{213}Bi alpha therapy.

The current ^{225}Ac supply can only support completion of Phase II clinical trials for treatment of AML. Completion of Phase III trials and deployment of the drug should it become FDA approved will require a significant expansion in production (Table 3.A.2 and Sidebar 4.1). Phase III clinical trials will require enough ^{225}Ac to be produced to support independent treatments at least three different institutions. A new pathway for production of ^{225}Ac via target irradiation will require a $1 0M infrastructure investment. Another option is to reconsider extracting ^{229}Th from the ^{233}U discussed in Chapter 8. However Congress has mandated the disposal of the 233U and major efforts are underway to carry out this mandate. If it were possible, initiating expansion of the production of ^{225}Ac from extracting ^{229}Th from ^{233}U might require $3.5M. Ramping up to 500 mCi generation of ^{225}Ac every two months to treat approximately 300 patients per year could occur in stages by adding additional ^{229}Th to the capability each year until the 500 mCi capacity is reached in FY21. It is estimated that the cost would be about $20M through FY21 in FY09 dollars. The need for this type of treatment could be very large, as there are 2000 people in the U.S. diagnosed with AML at any given time. The current treatment protocol calls for the patients to be first treated by standard techniques and if successful in putting the cancer into remission, when the cancer returns then the patient can be entered into the targeted alpha therapy clinical trials. Patients who do not respond to the standard treatment generally decline rapidly. This is a case where following successful trials, major new isotope production capabilities would be required for general use.

Los Alamos National Laboratory

The isotope production facility (IPF) at Los Alamos depends on a 100 MeV proton beam from the LANSCE facility utilized in a target irradiation facility, with nearby hot cells for separation of produced radioisotopes. This target irradiation facility allows for simultaneous multiple isotope production at different energies. Because of these capabilities, Los Alamos carries a large commercial isotope production role for the isotope program. The major isotopes produced and sold are summarized in Table 11.6, which illustrates, for example, that the quantity of ^{82}Sr (25 day half life) produced has grown from 28K mCi in FY05 to 66K mCi in FY08. ^{82}Sr treats thousands of patients per year. Around 100 mCi of ^{82}Sr is used to manufacture a generator, which provides four weeks of patient doses impacting 240 patients per generator. This ^{82}Sr production allows for PET imaging of the heart to detect myocardial infarctions, an effective life-saving diagnostic tool. Nearly 400 cardiac patients are diagnosed daily, so the need is great. ^{68}Ge is used for calibrating PET scanners in clinical use. There are more than 800 PET centers in the U.S. and over 1500 worldwide. Los Alamos and Brookhaven are the primary suppliers of ^{68}Ge and have coordinated programs to meet the increasing demand. In FY08, BNL produced approximately 42000 mCi of ^{82}Sr and 4600 mCi of ^{68}Ge.

Table 11.6. Top isotopes produced and sold at LANL over a five-year period

Isotope	Producer	FY05	FY06	FY07	FY08	FY09 (estimate)
Sr-82 (mCi)	IPF & INR	28574	540009.5	88673.5	66023	49809
Ge-68 (mCi)	IPF & iThemba	5121	8052	6358	10432	6591
Na-22 (mCi)	IPF		632		1485	
As-73 (mCi)	IPF		61	583		1323
Y-88 (mCi)	IPF		309.1		1266	
Gd-148 (mCi)	800 MeV Target				0.113	
Rb-83 (mCi)	By Product				6.2	
Be-7 (mCi)						6.5
Total Targets Processed		42	72	48	46	17
Total Product Shipments		83	164	117	117	79
Generators		490	513	821	715	360
Generator Shipments		28	25	46	23	10

Table 11.7. LANL FY08 budget and the FTEs supported

IPF Ops. Total	$3M	LANL Hot Cells Ops. Total	$4.3M
Salaries	$2.2M	Salaries	$3.4M
M&S	$600K	M&S	$700K
Other	$200K	Other	$200K
Workforce	7 (1 PhD, 6 others)	Workforce	13.0 (2.6 PhD, 10.4 others)

The appropriated funding for the Los Alamos isotope program is $4.64M, which supports 12 FTEs. The combined funding from sales and appropriation is ~$7.3M per year, which includes some preventive maintenance costs ($800K/yr) but does not include money for major infrastructure projects funded sporadically as upgrades or refurbishments. These funds and the associated FTEs (20 in total) break out as shown in Table 11.7 for FY08.

The laboratory is concerned that long-term sustainability is difficult in this program with the current business model and associated funding levels. Some of the challenges are listed below:

- The constant emphasis is on meeting immediate production needs; production fluctuations impact planning and staffing; shifting some of the FTEs funded on sales revenue to the program appropriation would help smooth these fluctuations and lead to a more sustainable program.
- The current focus is on capability maintenance via people retention and small short-term fixes to keep production on track; the laboratory meets changing production requirements as the first priority.
- Infrastructure age and maintenance deferred to meet production goals will eventually halt production if these problems are not addressed.

LANL sees large-scale refurbishments are needed to ensure that production operations continue in order to meet customer demand. These include

- beam window replacement at IPF,
- TA-48 Hot Cell train component replacement and modernization,
- electrical upgrade of TA-48 Hot Cells,
- manipulator replacement,
- TA-48 Hot Cell Window refurbishment,
- replacement of H^+ injector at accelerator - a long term need that would cost $2 to 10M, and
- backups for single point failures in the IPF control system and target loading system.

Cost estimates for the near-term infrastructure needs were estimated to total about 5.4M. Funding from the *American Recovery and Reinvestment Act* has provided $1.3M this year for some of these needs.

Los Alamos presented options for the capabilities to expand its production of isotopes, given funds and people to develop new processes. These include the following:

- Increase IPF and TA-48 utilization.
- Install an alpha emitter production capability at IPF and the CMR.
- Finish installation of ^{241}Am production capability at TA-55.
- Recover isotopes from existing inventory at TA-55.
- Utilize LANL's Stable Isotope Resource to provide labeled compounds as needed.
- Utilize the many other user facilities at LANSCE to perform parasitic irradiations for small-scale production of R&D isotopes.
- Install an isotope production capability at the Material Test Station.

Isotopes for the Nation's Future: A Long Range Plan

For example, it is possible to optimize an existing process (EXCEL) to mine [241] Am using an additional process (CLEAR - Chloride Line Extraction And Recovery). This CLEAR process is 80% installed and would require additional investment to complete. The fully developed process would yield an estimated ~500 g/year of [241] Am after production starts. A total of about $7M is needed to finish installation and to produce the first 250 g of the product, plus $2-3M per year of operating costs thereafter.

Brookhaven National Laboratory

The Brookhaven Linac Isotope Producer (BLIP) was the world's first facility to seriously exploit the isotope production capabilities of a high-energy proton accelerator. The use of higher-energy particles allows the use of relatively thick targets, where the large number of target nuclei can compensate for the generally smaller nuclear reaction cross sections compared to low-energy reactions.

BLIP was built in 1972 and utilizes the excess beam capacity of the 200-MeV proton Linac that injects into larger synchrotrons at BNL. A 30 m transport line delivers the protons to a shielded target area for radioisotope production. Work is done on process and target development to improve the quality and yield. The isotopes currently produced are shown in Table 11.8 and include $^{82}Sr/^{82}Rb$ for human heart scans with PET, ^{68}Ge for calibration of PET devices, and ^{65}Zn tracers for metabolic or environmental studies. Radioisotope research and development is being performed on ^{67}Cu for cancer therapy applications and ^{86}Y for cancer imaging. BLIP can operate parasitically (at lower shared operating costs) when the nuclear physics program at RHIC is using proton beams. Dedicated running with isotope production as the primary users is also available at higher operating costs.

Table 11.8. Major isotopes produced at BLIP at present

Isotope	# Shipped/Yr	Half-Life	Typical Application
Be-7	<1	53.3 d	γ source
Zn-65	~25	244 d	Zn tracer
Cu-67	1	61.9 h	Radioimmunotherapy
Ge-68/Ga68	15	271 d/68 m	PET calibration
As-73	2-3	80.3 d	As environmental tracer
Sr-82/Rb-82	20	5.4 d/75 s	PET studies of heart
Y-88	2-3	106.6 d	Y-90 tracer, γ source
Tc-95m	<1	61 d	Tc tracer

Appropriated funding for BLIP in FY09 is $3.47M; this includes 5.8 FTE staff labor, general maintenance of BLIP ($500K), and $2.95M for the Target Processing Lab (hot cells, radiochemistry laboratories, instrument rooms, liquid and solid radwaste storage capacity). Other programmatic costs are accelerator operations with beam, isotope processing (target fabrication, irradiation, hot cell isotope separation chemistry, assay, quality assurance, dispensing, shipping, waste disposal, financial reporting, etc.), and isotope development. In FY08 BLIP operated for six months with irradiation costs of $1.62M, isotope production costs of $266K (1.4 FTE and ~$66K materials), and R&D efforts costs of $193K (1.3 FTE). The maintenance, regulatory compliance, programmatic effort, research effort and work for others funding supported nine FTE total core staff in the group.

BNL presented the current infrastructure needs totaling $644K that include the following (of which $225K was provided in ARRA):

- Upgrade of ventilation system for ANSI/EPA compliance.
- Installation of a laser based proton beam energy sensor.
- Replacement of distilled water system.
- Replacement of hot cell acid vapor neutralization system.
- Replacement of multi-wire beam profile monitor.
- Installation of two vertical steering magnets into beam line.
- Purchase and installation of an inductively coupled plasma mass spectrometer (ICP-MS) to replace an old ICP optical emission spectrometer used for elemental assay of isotopic products.

BNL identified these future needs for longer-term viability of the facility:

- Replacing the BLIP groundwater cap and ramp.
- Accelerating research on ^{67}Cu and ^{86}Y, now under development.
- Upgrading the Linac to attain good product purity for ^{86}Y and allow low-energy operation by replacing existing beam loss monitors.

To improve isotope availability, there is a request to operate the Linac and BLIP for a longer period each year, up to a maximum of 9.5-10 months per year. This increase would require the use of hot cells for isotope processing to continue year round. Without any general off time for hot cell maintenance, a rotating schedule to remove hot cells from use would be implemented for maintenance and cleaning; radioactive waste disposal activities, which are now concentrated in downtimes, would, instead, have to be done continuously. An addition of 1.3 FTE would be required to support more infrastructure maintenance (due to less down-time) that would be needed. The existing menu of isotopes would continue on a longer production schedule.

In the long term, there is a proposal to expand the R&D effort to accelerate introduction of new or improved isotopes to the user community. The present four year cycle time is probably too slow for responding to national needs. To achieve this expanded research and development effort, a 2.2 FTE increase in research staff would be needed, maintenance costs are estimated to increase to $650K, hot cell maintenance cost to increase to $3.1 M, isotope processing costs to increase to $440K, the R&D effort would require $640K per year, resulting in total program annual costs of $7.9-9.0M per year.

BNL also proposed a response to needs for year-round delivery, higher capacity, and more research and workforce development capabilities through replacing the use of the BLIP linac with a stand-alone 70 MeV cyclotron, the Cyclotron Isotope Research Center. The capital cost was estimated at $38M (FY09$). Stony Brook University would be a university partner in the education and training plan. The increase in operating costs accompanying the increase in availability from 17 to 50 weeks per year were estimated to be $1-2M.

Proposed Changes in Isotope Program Operation

Analysis of the operations of the three major national laboratories contributing heavily to the isotope program reveals similar needs relating to people supported by the appropriated budget (compared to sales revenue), funds required for infrastructure maintenance and improvement, and the importance of research and development. It is important to consider implementation of the following changes to the isotope program.

- Fund more of isotope program staff on appropriations, in order to have a more stable operation.
- Increase yearly funding of infrastructure improvements.
- As discussed in Chapter 9, fund an R&D component for developing new techniques for radioactive and stable isotope production, in part from sales revenues, realizing that there will be fluctuation in these expenditures. This would include development of new isotope production, techniques for target fabrication and processing, source fabrication, and improving transportation options within supply chain.

The subcommittee viewed appropriate levels for these changes in an optimum budget (to be discussed in the next chapter) would be to work towards a 10% (of total appropriated plus sales resources) re-investment strategy in infrastructure improvement, a similar amount in research, and towards a ~25% increase in staffing on appropriated funding.

Pricing of Isotopes

An issue of critical importance to the research community is the price they must pay for the isotopes. There are examples of research directions not pursued or decreased in scope because of the unavailability of isotopes at what the research community or the responsible research funding agency deems 'reasonable costs.' The direction to charge full cost of production or the cost of isotope replacement has led to generally increased prices of isotopes. The availability of isotopes from abroad has not solved this problem, as foreign suppliers often price their isotopes a small increment below the DOE prices. It is important for the isotope program to examine devising a policy that would result in reduction in the prices of isotopes that are important for research. Of course, this is predicated on a process that

- Understands the cost of producing isotopes for research and/or commercial sales,
- Decides which isotope uses are research in nature, and
- Develops logical criteria to charge research customers in a responsible fashion that encourages important research.

A *unit cost* can be calculated and defined for a radioisotope that is produced in a batch mode. While this isotope is used for research purposes, e.g., to develop a new diagnostic or treatment process for patients, program leaders could decide what reduced cost can be charged while the quantities needed are small in this research phase. However, as the research phase succeeds and the process proceeds through phase II and into phase III clinical trials, the subsidy of the costs to achieve a reduced research price would need to end as large volumes of the radioisotope are needed. Stable isotopes present another set of pricing issues since some may be newly produced while others were produced long ago.

It is important to devise the strategy and criteria for defining certain isotopes uses as *research* in nature. Possibilities include identifying as research isotopes those isotopes that are among the following categories:

- Specified in national studies, e.g. recent reports of the National Academy of Sciences, DOE-NIH working groups, or this NSAC report.
- Defined by customer requests or Special Nuclear Material surveys.
- Alpha-emitting radionuclides as listed in Appendix 7, e.g. ^{225}Ac and ^{211}At.
- Isotopes for therapy and diagnostics as listed in Appendix 7, ^{67}Cu, ^{86}Y, ^{177}Lu.
- Heavy elements as listed in Appendix 7, e.g.. Cf, Ra, Bk.
- Related to focused study and R&D on new or increased production of He-3.
- Important for reestablishing a domestic supply of other isotopes.

Final selection could be based on availability of facility, capacity, cost of production, and quantities. Production of certain isotopes is limited to special sites, e.g. for transuranics, or the need for proton energies greater then 45MeV, or alpha-beams. Other isotopes would be produced based on peer-review results.

The question of whether an isotope can be defined both as research and as commercial in its pricing based on its use needs to be reexamined. One can certainly envision situations where this is logical and where research pricing is maintained even when the isotope is priced at full cost for commercial buyers. It may be possible to align the definition of research isotopes with the current Office of Sciences approach for user facilities. In this model, for-profit and proprietary R&D use would result in full payment of the cost for facility access, whereas non-profit companies and researchers that are not competing with commercial use are subsidized (if not provided that service at no cost). Under the current procedure, a company can in principle request, for example, 500 mCi of ^{225}Ac for the purposes of sale and get the same price as a federally funded researcher supporting a clinical trial for the same isotope, and perhaps even exhaust the available supply.

In summary, the subcommittee believes isotope program leaders should pursue a pricing model that distinguishes between research and commercial uses. The model should incorporate the ability to maintain research pricing in certain cases even when the isotope is priced at full cost for commercial buyers. The justification would seem appropriate in defined areas of national research need.

Program Coordination and Outside Input

The isotope program is a very important national resource that serves many users and customers, benefits R&D in a number of disciplines, is crucial for the biomedical community, and depends on the production of isotopic material at different sites. It is essential that emphasis be put on coordinating this vast program across many facilities and laboratories that produce isotopes (DOE and others) and on getting consistent and informed input from users from the various communities.

Program leaders should devote special attention to the coordination of production capability over the broad program at national laboratories as well as universities. Of course, this requires a concerted effort to anticipate demand in isotopes as much as possible. Elements of this program coordination include the following:

- Predict demand over a multi-year period, especially relating to the large-volume isotopes.
- Anticipate swings in the global market for isotopes, as rapid changes abroad affect the pricing and availability of isotopes for domestic users.
- Coordinate production facilities at national laboratories and at universities. LANL and BNL already coordinate to attain year-round availability of longer-lived isotopes that both can produce, e.g., ^{82}Sr, ^{68}Ge, ^{88}Y, and ^{7}Be. This coordination extends to sharing procedures, staff, shipping containers, and the isotopes themselves.
- Maintain close communication with other federal agencies that depend on supply of isotopes. The committee has been impressed by the wide demand for isotopes in the programs of various federal agencies. It is important to monitor those needs since sometimes the demands for certain isotopes are large in quantity.
- Cooperate with foreign suppliers since self-sufficiency in domestic production is unlikely. Arrangements with foreign accelerators in Russia and France to supplement domestic supplies have already been implemented by DOE. But this cannot be extended to short-lived isotopes because of the logistics of shipping around the world. In practice it becomes difficult to do this for half lives less than 10 days or so.

The community that uses and depends on isotopes is very broad and the needs are often very diverse. In view of the breadth and diversity of the demand, it is important for DOE to regularly seek advice from outside advisory bodies of different types. BNL and ORNL at one time had user groups to bring outside input to the isotope program. However, these ceased operation. At this crucial stage of revamping the isotope program, it is important to form and pay attention to various levels of outside advisory bodies:

- Initiate a users committee to aid in establishing R&D priorities and to provide input on scientific issues of strategic importance.
- Constitute proposal review committees to (a) evaluate and prioritize individual research proposals based on scientific merit, feasibility, capability of the experimental group, and availability of resources, and (b) to advise on definition of research isotopes and when they become commercial.
- Organize and facilitate activities of an executive-level advisory body to provide longer- term programmatic advice.

As the isotope program is re-formed and re-organized towards an emphasis on isotope production R&D, customer satisfaction, education, and outreach, it will be essential to build the information technology tools necessary for this new mission. A modernized web page is a necessity, both to dramatically enhance the program's online presence and also to host custom software tools that could facilitate R&D, collaboration building, information sharing, and program management. Such tools would enable the creation of an online community that is interested in isotope R&D and utilization, and would set a new paradigm for communication between the new program and its stakeholders, at a level never present in past isotope work. Information about the availability of specific isotopes, the form of the isotope desired, production techniques and yields, and their utilization are among some of the information that would be extremely useful to the research and end-user communities.

This combination of knowledge of isotope demand and coordination would also place the program and its advisers in the position to help identify critical radioisotopes that could be viewed as strategic. This would help alert the government and communities of potential issues, even for isotopes beyond the responsibility of the isotope program, and encourage solutions based on coherent long-range planning.

Isotope Transportation

The safe, efficient, and cost-effective shipment of radioisotopes is a crucial element of the isotope program. An investment of funds is needed to develop a Type B cask and to make specific improvements to dated Type A container designs that would benefit all laboratories producing radioisotopes.

Packages for shipment of radioisotopes are labeled as Type A and Type B based on the level of radioactivity of the specific isotope being shipped. Type A designs are tested to meet normal transport conditions. Testing and certification can be performed by any organization as long as the proper testing is well documented. The cost to design and test Type A packages is relatively inexpensive, $50K to $ 100K. Such packages can consist of various forms, from fiberboard boxes to large shielded containers. They can be approved for liquids or solids, and higher activity levels of solids can be shipped in Type A packages when encapsulated in welded capsules (special form capsules). The current non-returnable Type A package designs used for liquids are limited to small volumes (<25 ml) and/or have limited shielding and have been in service for over 40 years. There are some returnable Type A packages that are commercially available, however, the shielding tends to be insufficient for many of the high-activity liquid shipments, and customers tend to prefer the non-returnable packages. The advantage of a Type A package with a higher liquid volume capacity and more shielding are the logistical efficiencies of a smaller number of shipments for those shipments which must arrive on time due to short half-lives and/or clinical treatment schedules for patients. By limiting the radiation readings to less than 3 mR/hr at one meter, one can ship medical research isotopes via commercial passenger airlines, which makes more flights available and results in fewer shipping delays. At present, the standard W/Re generators and bulk solutions of W/Re must be shipped only by cargo aircraft, resulting in potential delays during transport, especially for international shipments.

Current Type A packages for ^{252}Cf only provide shielding for up to 3.7 mg, which is not compatible with at least one of the major customers of this radioisotope. With additional shielding, one could increase the quantity shipped up to the Type A limit of 5 mg. The current W/Re generator was developed as a modified design of an existing non-returnable Type A package, since this was the most economical approach. Design features that could have enhanced the usefulness of the generator were not pursued.

There are three areas of needed improvements for Type A packages:

- ^{252}Cf containers. A new design is needed to provide enough shielding for up to 5 mg of ^{252}Cf. It must be low in weight (<5,000 lbs) to avoid engineering tie downs and allow movement by normal commercial freight. It must have sufficient internal volume for a variety of welded special form capsules. And, it needs to be compatible with production facilities at DOE sites and with a customer's facilities, e.g., to be unloaded into their hot cell.
- General use solid/liquid transport container. This would offer large liquid volume capacity, have shielding capability for passenger aircraft transport, have low weight (<100 lbs), and be compatible with production and customers' facilities.
- A radioisotope generator design that addresses both the unique generator application features and the Type A qualification requirements would be more efficient and user friendly. For example, this would make the W/Re generator more attractive and potentially easier to transfer to the private sector.

Type B designs are tested to hypothetical accident conditions and are certified by the Nuclear Regulatory Commission. The process to design, test, and certify Type B containers is expensive. In October of 2008, the Department of Transportation discontinued certification on two widely used Type B packages, the 6M and 20WC specification designs (See Figure 11.7). These two general use packages limited shipments by two parameters; 100 watts and 200 mR/hr surface reading and were widely used by the DOE isotope program for a range of isotopes. Decertification of these designs has limited the program's transportation capabilities.

ORNL currently does not have a Type B cask for ^{252}Cf. This limits the program to 5 mg sources shipped in Type A packages, but key customers have historically used up to 30 mg sources for their applications. Additionally, with the expected growth in the ^{252}Cf commercial source market, the capability to ship Type B quantities of bulk ^{252}Cf would improve the efficiency of the supply chain. Otherwise, multiple shipments of smaller quantities of ^{252}Cf are required which increases handling and personnel exposure.

One or more new Type B shipping cask designs need to be developed, tested, and certified and/or purchased if commercially available. They should have the following features:

- A large Type B cask design for high-activity shipments (kilocuries) of ^{60}Co, ^{252}Cf, or other isotopes with intense gamma-ray emission.
- A small Type B cask design (<500 lbs) for air transport (e.g., use for ^{192}Ir, ^{75}Se, ^{249}Bk or other products).
- A cask that is compatible with all DOE reactor/accelerator facilities and associated hot cell facilities.
- Drain capability to allow for shipment directly from a pool reactor, if hot cell processing is not needed.
- A cask approved for a wide range of isotopes in either normal (non-welded capsule) or special form (welded capsules).
- A cask that incorporates "next generation" security safeguards (e.g., secure closure, built- in tracking capability, etc.).

In many cases, internal containment requirements for Type B containers or Type A containers specify the use of inner special form welded capsules in their certifications. The capacity to perform this remote capsule welding throughout the DOE isotope program facilities is necessary for an efficient transportation program.

Figure 11.7. A SRP 25-ton Target Tube Cask Type B shipping container that is used for onsite shipments but has been decertified for off-site shipments.

An infrastructure investment of several millions of dollars is needed to develop and certify such a Type B shipping cask. While the primary needs for these investments are currently found at ORNL, a coherent focus on transportation issues can benefit all parts of the DOE isotope program involved in the production and shipping of radioisotopes, and perhaps play an important role in increased coordination and networking with other facilities. This is one high priority use for part of the sustained increase in infrastructure that is part of the optimum budget scenario presented in the next chapter.

Summary Recommendations

Maintain a continuous dialogue with all interested federal agencies and commercial isotope customers to forecast and match realistic isotope demand and achievable production capabilities.

For the isotope program to be efficient and effective for the nation, it is essential that isotope needs be accurately forecast. The DOE-NIH interagency working group is an excellent start for this type of communications in a critical area of isotope production and use.

Coordinate production capabilities and supporting research to facilitate networking among existing DOE, commercial, and academic facilities.

Support a sustained research program in the base budget to enhance the capabilities of the isotope program in the production and supply of isotopes generated from reactors, accelerators, and separators.

Devise processes for the isotope program to better communicate with users, researchers, customers, students, and the public and to seek advice from experts:

- Initiate a users group to increase communication between isotope program management and users on issues of availability, schedules, priorities, and research.
- Form expert panels as needed to give advice on issues such as definition of isotopes as research or commercial in primary usage, new production methods, and needed actions when demand exceeds supply.
- Modernize the web presence for the isotope program to give users an easier way both to learn about properties, availability, production methods, and services, and also to have access to interactive tools that help customers plan purchases and use, researchers to share information and form collaborations, and students and the general public to learn about the important uses of isotopes.

Encourage the use of isotopes for research through reliable availability at affordable prices.

Increase the robustness and agility of isotope transportation both nationally and internationally.

- Identify and prioritize transportation needs through establishing a transportation working group.
- Initiate a collaborative effort to develop and resolve the priority issues (i.e., certification of transportation casks).

12. BUDGET SCENARIOS

The budget projections were based on the FY09 President's request of $1 9.9M. With the guidance from the Office of Nuclear Physics, the constant effort scenario extended this total budget forward in constant FY09 dollars until 2018. In the 2009 President's request, $3.1M was provided to the Research Isotope Development and Production Subprogram to support isotope production and research and development activities of commercially-unavailable research isotopes. As discussed in more detail in Chapter 11, the FY09 Omnibus bill appropriated $24.9M to the isotope program, with $4.9M provided for the Research Isotope Development and Production Subprogram. Also, under the American Recovery and Reinvestment Act (ARRA), the Department of Energy allocated $14.617M to the isotope program for highly needed infrastructure investments and for isotope production that will improve America's competitiveness by investing in scientific research at laboratories and universities and contributing to training the scientific and technical workforce the U.S. needs. The detailed funding profile for the ARRA funds is still being defined, and an average profile was assumed over the next three years. These increments are included in the budget projections, but the constant effort level of appropriations from FY10 to FY18 is still assumed to be $19.9M (FY09).

The optimum budget projection provides a base of funds for research and development (at ~1 0% of the total operating budget, that is appropriations plus sales minus construction), additional funds to maintain infrastructure at the current level of performance (at ~10% of operations), provides stable funding for key skilled individuals so that capabilities will not be lost with significant downward fluctuations in sales revenue (~25% increase in manpower on appropriations minus construction), and identifies funds for workforce development (~5% of appropriations minus construction). The ARRA funds in 2009-2011 allow these increases to be phased in so the optimum levels are reached in 2012. Including these adjustments leads to an appropriated operations budget of about $25M in 2012, the same level as the FY09 appropriations.

Four initiatives are included with approximate resources required given in FY09$: 1) funds for proof-of-principle demonstration of the implementation of increased production of alpha- emitting isotopes for therapy assuming this remains a high priority for NIH research ($4M total from 2011-2013), 2) a new cyclotron facility ($40M total from 2011-2014), 3) a new electromagnetic isotope separation facility ($25M total from 2012-2015), and 4) in the longer term the start of a future initiative that could address a significant increase in demand as research opportunities expand into general use ($50M total starting in 2016). The nature and cost of such a facility will clearly depend on the evolution of future opportunities and the response of the private sector to commercial isotope opportunities. The schedules outlined for these new facilities are clearly aggressive. The subcommittee cannot be aware of the DOE FY20 11 budget request. The important take-away messages are the need to address alpha isotopes early, that the isotope separator has the highest priority for the major capital requests, but the need is not as immediate and such a facility might benefit from R&D and more deliberate planning. Therefore, in this scenario construction on the accelerator was started and completed earlier. The first three elements could be realized with an annual capital budget of about $1 5M per year for several years.

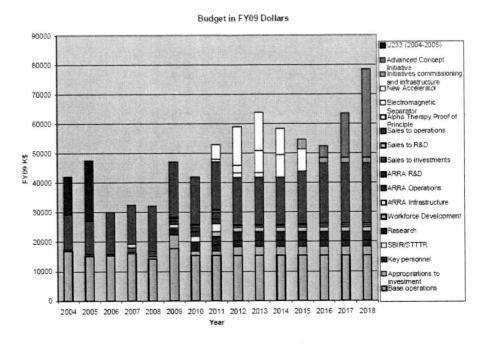

Figure 12.1. Proposed optimum budget in thousands of FY09 dollars.

The subcommittee did not have detailed costing information available for most of the initiatives considered, so the actual cost estimates should be considered qualitative, perhaps at the 30% level. However, NSACI could also see the possibility for substantial reductions in these costs by effective use of existing resources or public-private partnerships.

Since the subcommittee heard arguments that sales with the current capabilities would both increase or decrease beyond FY09, it arbitrarily assumed that sales would remain constant at the $17M (FY08) level until new capabilities came on-line. At that point there would be some commissioning costs and then sales were projected to cover the additional operating costs except for a small additional infrastructure maintenance cost. Figure 12.1 presents the optimum budget scenario in thousands of FY09 dollars.

Under a constant effort budget scenario, the subcommittee believes that it is still important to provide funds for research and development (~4% of total operations), invest in infrastructure (5% of operations), and provide stable funding for key skilled individuals (20% increase of manpower on appropriations minus construction), and workforce development (1.5% of appropriations). Given the constraint on the total funds, the subcommittee had to choose to redirect these funds from the Research Isotope Development and Production Subprogram. The additional investments in 2009 and the ARRA investments will allow the program to move forward from a more solid base for a few years. However, once this funding disappears, sustained constant effort level funding, while it does represent a needed increase from the 2004-2008 levels, will continue to place the infrastructure needs for research isotopes at risk. Figure 12.2 presents the constant effort budget scenario.

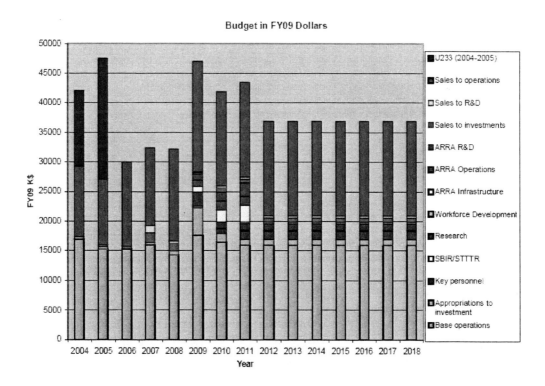

Figure 12.2. Constant effort budget in thousands of FY09 dollars.

For any budget levels at constant effort or above, the priorities presented here for the program should be clear: have continuing stable investments in R&D, workforce development, and infrastructure, and stabilize the funding for the highly skilled manpower required within the base program. However, without the new accelerator and new isotope separation capabilities, the nation will be without vital isotope production capabilities that are essential for research and advanced technology. A constant effort budget or below will force the nation to rely heavily on uncertain foreign sources of isotopes and certain isotopes will not be available reliably for research.

13. SUMMARY OF RECOMMENDATIONS FOR CHARGE 2

The major recommendations of the NSACI subcommittee in response to Charge 2 presented in the earlier chapters are summarized here. The recommendations are divided into three categories: I) Recommendations about the present program, II) Development of a highly skilled workforce for the future, and III) Major investments in production capacity to provide capabilities not available to the nation's current isotope program. The recommendations in the first category are listed in order of priority and the relative priorities of the recommendations in the 2nd and 3rd categories are discussed below.

The Present Program

I.1: Maintain a continuous dialogue with all interested federal agencies and commercial isotope customers to forecast and match realistic isotope demand and achievable production capabilities.

For the isotope program to be efficient and effective for the nation, it is essential that isotope needs be accurately forecast. The DOE-NIH interagency working group is an excellent start for this type of communications in a critical area of isotope production and use.

I.2: Coordinate production capabilities and supporting research to facilitate networking among existing DOE, commercial, and academic facilities.

In the short term, increased isotope production and the availability of new research isotopes require more effectively exploiting the available production facilities including resources outside those managed by the isotope program. This will require both research and development to standardize efficient production target technology and chemistry techniques and flexible funding mechanisms to direct production resources most effectively.

I.3: Support a sustained research program in the base budget to enhance the capabilities of the isotope program in the production and supply of isotopes generated from reactors, accelerators, and separators.

Research and development may significantly expand the production efficiency and capacity of the program. It is also an important path to expanding the skilled isotope production workforce and retaining the most creative people in the program.

I.4: Devise processes for the isotope program to better communicate with users, researchers, customers, students, and the public and to seek advice from experts.

- Initiate a users group to increase communication between isotope program management and users on issues of availability, schedules, priorities, and research.
- Form expert panels as needed to give advice on issues such as definition of isotopes as research or commercial in primary usage, new production methods, and needed actions when demand exceeds supply.
- Modernize the web presence for the isotope program to give users an easier way both to learn about properties, availability, production methods, and services, and also to have access to interactive tools that help customers plan purchases and use, researchers to share information and form collaborations, and students and the general public to learn about the important uses of isotopes.

I.5: Encourage the use of isotopes for research through reliable availability at affordable prices.

Many research applications, and especially medical trials, cannot proceed without a dependable source of isotopes. At the same time, DOE should reexamine its pricing policy for research isotopes to encourage U.S. leadership in isotope-based research.

I.6: Increase the robustness and agility of isotope transportation both nationally and internationally.

- Identify and prioritize transportation needs through establishing a transportation working group.
- Initiate a collaborative effort to develop and resolve the priority issues (i.e., certification of transportation casks).

Highly Trained Workforce for the Future

II: Invest in workforce development in a multipronged approach, reaching out to students, post-doctoral fellows, and faculty through professional training, curriculum development, and meeting/workshop participation.

The dwindling population of skilled workers in areas relating to isotope production and applications is a widely documented concern. This recommendation is focused on the needs of the IDPRA program, itself. The relative priority of this recommendation is comparable to that for a sustained R&D program, with which it is closely linked.

Major Investments in Production Capability

The present program is highly flexible and responsive to the needs of the nation. However, it lacks two major capacities that seriously limit its ability to fulfill its mission. The isotope program presently has no working facilities for the separation of a broad range of stable and long-lived isotopes. Each year it is depleting its unique stockpile of isotopes to the point where some are no longer available. Secondly, many radioactive isotopes by their very nature can be short-lived and cannot be stockpiled. The current program relies on accelerators and reactors whose primary missions are not isotope production; thus, it is not in a position to provide continuous access to many of the isotopes.

III.1: Construct and operate an electromagnetic isotope separator facility for stable and long-lived radioactive isotopes.

It is recommended that such a facility include several separators for a raw feedstock throughput of about 3 00-600 milliAmpere (10-20 mg/hr multiplied by the atomic weight and isotopic abundance of the isotope). This capacity will allow yearly sales stocks to be replaced and provide some capability for additional production of high-priority isotopes.

III.2: Construct and operate a variable-energy, high-current, multi-particle accelerator and supporting facilities that have the primary mission of isotope production.

The most cost-effective option to position the isotope program to ensure the continuous access to many of the radioactive isotopes required is for the program to operate a dedicated accelerator facility. Given the uncertainties in future demand, this facility should be capable of producing the broadest range of interesting isotopes. Based on the research and medical opportunities considered by the subcommittee, a 3 0-40 MeV maximum energy, variable energy, high-current, multi-particle cyclotron seems to be the best choice on which to base such a facility.

The subcommittee gives somewhat higher overall priority to the electromagnetic isotope separator as there is no U.S. replacement. However, a solution in this area is not needed as urgently as the new accelerator capability. Therefore, in the subcommittee's optimum budget scenario that includes both, the construction of the new accelerator starts a year earlier.

The implications of these recommendations are discussed in an optimal budget scenario and under a constant level of effort budget (taken to be the 2009 President's request of $19.9M). Given the recent investments in the isotope program, especially significant American Recovery and Reinvestment Act (ARRA) funding, constant effort funding will allow the program to move forward from a more solid base for a few years. Once this ARRA funding disappears, sustained constant effort level funding, while it does represent a needed increase from the 2004-2008 levels, will place the infrastructure needs for research isotopes at risk in the long term and will not allow the program to address either of the two major missing capacities. The subcommittee does not consider this to be a wise course for the future. The subcommittee recommends an optimum budget that reaches a sustained base operating funding of about $25M (FY09$) per year and also includes new capital funds of about $1 5M (FY09$) per year for several years to realize the needed new capacities.

In addition to its recommendations, the subcommittee noted one major concern:

The supply of ^{99}Mo, the isotope used to generate the radioactive isotope most frequently used in medical procedures, is of great concern. Recent disruptions in international supply demonstrate the vulnerability of the nation's health care system in this area. The nation must address this vulnerability. At the present time, the isotope program does not produce ^{99}Mo. With the non-proliferation issues associated with the transport and use of the highly-enriched uranium currently used for ^{99}Mo production, DOE/NNSA has the lead responsibility in this area and is actively investigating options for ^{99}Mo commercial production. The subcommittee chose to refrain at this time from inserting itself into the intense activity underway but reiterates the importance of the issue.

REFERENCES

[AA08] "Nuclear Weapons in 21st Century and U.S. National Security", Report from the AAAS, the APS, and the Center for Strategic and International Studies (2008).

[AB92] S. A. Abrams, P. D. Klein, V. R. Young, and D. M. Bier, "Letter of Concern Regarding a Possible Shortage of Separated Isotopes", J Nutr. 122, 2053 (1992).

[AH99] Ad Hoc Committee for the North American Society for the Study of Obesity. "Report on the Supply and Demand of Oxygen-18 Water", http://www.obesity.org/about/19990121.asp, January 21, 1999.

[AP08] "Readiness of the U.S. Nuclear Workforce for 21st Century Challenges", Report from the APS Panel on Public Affairs Committee on Energy & Environment (2008).

[AP09] "Nuclear Forensics – Role, State of the Art, Program Needs", Joint Working Group of the American Physical Society (APS) and the American Association for the Advancement of Science (AAAS) (2009).

[BE04] "Radiopharmaceutical Development and the Office of Science", Biological and Environmental Research Advisory Committee Subcommittee (2004).

[CE09] T. E. Cerling, G. Wittemyer, J. R. Ehleringer, C. H. Remien, and I. Douglas-Hamilton, "History of Animals Using Isotope Records (HAIR): a 6-Year Dietary History of One Family of African Elephants, Proc Natl. Acad. Sci. 106, 8093-8 100 (2009).

[CH09] Y. Cherel, V. Ridoux, J. Spitz, and P. Richard, "Stable Isotopes Document the Trophic Structure of a Deep-Sea Cephalopod Assemblage Including Giant Octopod and Giant Squid", Biol. Lett. 23, 364-367 (2009).

[CO38] W. E. Cohn and D. M. Greenberg, "Studies in Mineral Metabolism with the Use of Artificial Radioactive Isotopes: I. Absorption, Distribution and Excretion of Phosphorus", J Biol. Chem. 123, 185-198 (1938).

[DOE05] "Management of the Department's Isotope Program", DOE Office of Inspector General, Audit Report DOE/IG-0709 (2005).

[FA02] S. J. Fairweather-Tait and J. Dainty, "Use of Stable Isotopes to Assess the Bioavailability of Trace Elements: A Review", Food Addit. Contamin. 19 939-947 (2002).

[FA13] Kasimir Fajans, "Über eine Berziehung zwischen der Art einer radioaktiven Umwandlung und dem elektrochemischen Verhalten der betreffenden Radioelemente", Physicalische Zeitschrift 14, 131-136 (1913)

[GU9 1] K. Y. Guggenheim, "Rudolf Schoenheimer and the Concept of the Dynamic State of Body Constitutents", J Nutr. 121, 1701-1704 (1991).

[HA59] P. W. Hahn, W. F. Bale, E. O. Lawrence, and G. H. Whipple, "Radioactive Iron and Its Metabolism in Anemia. Its Absorption, Transportation and Utilization", J Exp. Med. 69, 739-753 (1939).

[IM95] "Isotopes for Medicine and Life Sciences", Institute of Medicine (1995).

[ITS09] United States International Trade Commission, "Stable and Radioactive Isotopes, Industry and Trade Summary", Office of Industries Publication ITS-01 (2009).

[KE01] E. P. Kennedy, "Hitler's Gift and the Era of Biosynthesis", J Biol. Chem. 276, 42619-42631 (2001).

[KO06] R. D. Koudelka et al., "Radioisotope Micropower Systems Using Thermo-photovoltaic Energy Conversion", AIP Conf. Proc. 813, 545-55 1 (2006).

[LY07] N. Lynnerup, "Mummies", Am. J Phys. Anthropol. Suppl. 45, 162-190 (2007).
[MC06] K. D. McKeegan *et al.*, Science 314, 1724 (2006).
[NA93] Y. Nagano *et al.*, J. Phys. Chem. 97, 6897-6901 (1993).
[NE99] "Forecast Future Demand for Medical Isotopes", Expert Panel Report to the DOE Office of Nuclear Energy (1999).
[NE00] "Final Report, Nuclear Energy Research Advisory Committee Subcommittee for Isotope Research and Production Planning", Nuclear Energy Research Advisory Committee Subcommittee (2000).
[NC08] "Report of the Meeting to Discuss Existing and Future Radionuclide Requirements for the National Cancer Institute, Expert Panel Report (2008).
[NO90] "Simple Thermochromatographic Separation of ^{67}Ga from Metallic Zinc Targets", A. F. Novgorodov *et al.*, Isotopes in Environmental and Health Studies 26, 118 (1990).
[NO08] J. Norenberg, P. Staples, R. Atcher, R. Tribble, J. Faught, and L. Riedinger, "Workshop on the Nation's Needs for Isotopes: Present and Future", http://www.sc.doe.gov/henp/np/program/docs/Workshop%20Report_final.pdf, (2008).
[NR07] "Advancing Nuclear Medicine Through Innovation, National Research Council Committee on State of the Science of Nuclear Medicine (2007).
[NR08] "Radiation Source Use and Replacement", National Research Council Committee on Radiation Source Use and Replacement (2008).
[NR09] "Medical Isotope Production Without Highly Enriched Uranium", National Research Council Committee on Medical Isotope Production Without Highly Enriched Uranium (2009).
[NS04] "Education in Nuclear Science: A Status Report and Recommendations for the Beginning of the 21st Century", NSAC Sub-Committee on Education, http://www.sc.doe.gov/henp/np/nsac/docs/NSAC_CR_education_report_final.pdf, (2004).
[NS07] "The Frontiers of Nuclear Science. A Long Range Plan" Nuclear Science Advisory Committee. http://www.sc.doe.gov/henp/np/nsac/docs/Nuclear-Science.High-Res.pdf, (2007).
[NS07A] "A Vision for Nuclear Science Education and Outreach", White Paper for 2007 Nuclear Physics LRP (2007).
[NS09] "Compelling Research Opportunities Using Isotopes", Report of the Nuclear Science Advisory Committee Isotopes Subcommittee, http://www.sc.doe.gov/henp/np/nsac/docs/NSAC_Final_Report_Charge1%20(3).pdf, (2009).
[QA82] S. M. Qaim, "Nuclear Data Relevant to Cyclotron Produced Short-Lived Medical Radioisotopes", Radiochim. Acta. 30(3), 147-162 (1982).
[QA01] S. M. Qaim, "Nuclear Data Relevant to the Production and Application of Diagnostic Radionuclides", Radiochim. Acta. 89, 223-232 (2001).
[QA04] S. M. Qaim, "Use of Cyclotrons in Medicine", Radiation Physics and Chemistry 71, 9 17-926 (2004).
[RA09] R. Ramos, J. Gonzalez-Solis, and X. Ruiz, "Linking Isotopic and Migratory Patterns in a Pelagic Seabird", Oecologia. 160, 97-105 (2009).
[RI05] M. J. Rivard *et al.*, "The U. S. National Isotope Program: Current Status and Strategy for Future Success, American Nuclear Society Special Committee on Isotope Assurance", Appl. Rad. Isotopes 63, 157 (2005).
[RU89] T. J. Ruth, B. D. Pate, R. Robertson, and J. K. Porter, "Radionuclide Production for the Biosciences", Nucl. Med. Biol. 16, 325-336 (1989).

[SC35A] R. Schoenheimer, "Deuterium as an Indicator in the Study of Intermediary Metabolism", Science 82, 156-157 (1935).

[SC35B] R. Shoenheimer and D. Rittenberg, "Deuterium as an Indicator in the Study of Intermediary Metabolism I", J Biol. Chem. 111, 163-168 (1935).

[SC40] R. Schoenheimer and D. Rittenberg, "The Study of Intermediary Metabolism with the Aid of Isotopes", Physiol. Rev. 20, 218-248 (1940).

[SC42] R. Shoenheimer, "The Dynamic State of Body Constituents", Harvard University Press. Cambridge (1942).

[SI02] R. D. Simoni, R. L. Hill, and M. Vaughan, "The Use of Tracers to Study Intermediary Metabolism: Rudolf Schoenheimer", J Biol. Chem. 277, e31-e33 (2002).

[SN05] "National Radionuclide Production Enhancement Program: Meeting Our Nation's Needs for Radionuclides", Society of Nuclear Medicine National Radionuclide Production Enhancement Task Force (2005).

[SO10] F. Soddy, "Radioactivity," Chemical Society Annual Reports 7, 256-286 (1910).

[SO13] F. Soddy, Chemical Society Annual Reports 10, 262-288 (1913).

[ST08] S. Stürup, H. Rüsz Hansen, and B. Gammelgaard, "Application of Enriched Stable Isotopes as Tracers in Biological Systems: A Critical Review", Anal. Bioanal. Chem. 390, 541-554 (2008).

[TU06] J. R. Turnlund, "Mineral Bioavailability and Metabolism Determined by Using Stable Isotope Tracers", J Animal Sci. 84, E73-E78 (2006).

[TR08] Task Force on Alternatives for Medical Isotope Production, "Making Medical Isotopes," http://admin.triumf.ca/facility/5yp/comm/Report-vPREPUB.pdf (2008).

[WA91] E. Wada, H. Mizutani, and M. Minagawa, "The Use of Stable Isotopes for Food Web Analysis". Crit. Rev. Food Sci. Nutr. 30, 361-71 (1991).

APPENDIX 1: THE NSAC CHARGE

August 8, 2008
Professor Robert E. Tribble
Chair, DOE/NSF Nuclear Science Advisory Committee
Cyclotron Institute
Texas A&M University
College Station, TX 77843

Dear Professor Tribble:

The Fiscal Year (FY) 2009 President's Request Budget proposes to transfer the Isotope Production Program from the Department of Energy (DOE) Office of Nuclear Energy to the Office of Science's Office of Nuclear Physics, and rename it the Isotope Production and Applications program. In preparation for this transfer, this letter requests that the Nuclear Science Advisory Committee (NSAC) establish a standing committee, the NSAC Isotope (NSACI) subcommittee, to advise the DOE Office of Nuclear Physics on specific questions concerning the National Isotope Production and Applications (NIPA) Program. NSACI will be constituted for a period of two years as a subcommittee of NSAC. It will report to the DOE

162   Nuclear Science Advisory Committee Isotopes Subcommittee

through NSAC who will consider its recommendations for approval and transmittal to the DOE.

Stable and radioactive isotopes play an important role in basic research and applied programs, and are vital to the mission of many Federal agencies. Hundreds of applications in medicine, industry, national security, defense and research depend on isotopes as essential components. Over the years, individual communities and Federal agencies have conducted their own studies, identifying their needs in terms of isotope production and availability. Most recently, the DOE Office of Nuclear Energy and the Office of Science's Office of Nuclear Physics organized a workshop to bring together stakeholders (users and producers) from the different communities and disciplines to discuss the Nation's current and future needs for stable and radioactive isotopes, as well as technical hurdles and viable options for improving the availability of those isotopes.

The next step is to establish the priority of research isotope production and development, and the formation of a strategic plan for the NIPA Program, in which we expect NSACI to play a vital role. The NIPA's products and services are sold world-wide both to researchers and commercial organizations. The NIPA produces isotopes only where there is no U.S. private sector capability or when other production capacity is insufficient to meet U.S. needs. Commercial isotope production is on a full-cost recovery basis. The following two charges are posed to the NSAC subcommittee:

Charge 1

As part of the NIPA Program, the FY 2009 President's Request includes $3,090,000 for the technical development and production of critical isotopes needed by the broad U.S. community for research purposes.

NSACI is requested to consider broad community input regarding how research isotopes are used and to identify compelling research opportunities using isotopes.

The subcommittee's response to this charge should include the identification and prioritization of the research opportunities; identification of the stable and radioactive isotopes that are needed to realize these opportunities, including estimated quantity and purity; technical options for producing each isotope; and the research and development efforts associated with the production of the isotope. Timely recommendations from NSACI will be important in order to initiate this program in FY 2009; for this reason an interim report is requested by January 31, 2009, and a final report by April 1, 2009.

Charge 2

The NIPA Program provides the facilities and capabilities for the production of research and commercial stable and radioactive isotopes, the scientific and technical staff associated with general isotope development and production, and a supply of critical isotopes to address the needs of the Nation. NSACI is requested to conduct a study of the opportunities and priorities for ensuring a robust national program in isotope production and development, and to recommend a long-term strategic plan that will provide a framework for a coordinated implementation of the NIPA Program over the next decade.

The strategic plan should articulate the scope, the current status and impact of the NIPA Program on the isotope needs of the Nation, and scientific and technical challenges of isotope production today in meeting the projected national needs. It should identify and prioritize the most compelling opportunities for the U.S. program to pursue over the next decade, and articulate their impact.

A coordinated national strategy for the use of existing and planned capabilities, both domestic and international, and the rationale and priority for new investments should be articulated under a constant level of effort budget, and then an optimal budget. To be most helpful, the plan should indicate what resources would be required, including construction of new facilities, to sustain a domestic supply of critical isotopes for the United States, and review the impacts and associated priorities if the funding available is at a constant level of effort (FY 2009 President's Request Budget) into the out-years (FY 2009 – FY 2018). Investments in new capabilities dedicated for commercial isotope production should be considered, identified and prioritized, but should be kept separate from the strategic exercises focused on the remainder of the NIPA Program.

An important aspect of the plan should be the consideration of the robustness of current isotope production operations within the NIPA program, in terms of technical capabilities and infrastructure, research and development of production techniques of research and commercial isotopes, support for production of research isotopes, and current levels of scientific and technical staff supported by the NIPA Program. We request that you submit an interim report containing the essential components of NSACI's recommendation to the DOE by April 1, 2009, and followed by a final report by July 31, 2009.

These reports provide an excellent opportunity for the Nuclear Physics program to inform the public about an important new facet of its role in the everyday life of citizens, in addition to the role of performing fundamental research. We appreciate NSAC's willingness to take on this important task, and look forward to receiving these vital reports.

Sincerely,
Jehanne Simon-Gillo
Acting Associate Director of the Office of Science
for Nuclear Physics

SC-26. 1 :EAHenry:cls: 903-3614:08/08/08 :q:\NSAC Directory\2008\NSAC Isotope Charge letter_Final.doc

SC-26.1 SC-26

E. A. Henry   J. Simon-Gillo
08/ /08   08/ /08

APPENDIX 2: MEMBERSHIP OF NSAC ISOTOPE SUBCOMMITTEE

Ercan Alp Ph.D.
Argonne National Laboratory
eea@aps.anl.gov

Ani Aprahamian Ph.D. (co-chair)
University of Notre Dame
aapraham@nd.edu

Robert W. Atcher Ph.D.
Los Alamos National Laboratory
ratcher@lanl.gov

Kelly J. Beierschmitt Ph.D.
Oak Ridge National Laboratory
beierschmitt@ornl.gov

Dennis Bier M.D.
Baylor College of Medicine
dbier@bcm.tmc.edu
Roy W. Brown
Council on Radionuclides and
Radiopharmaceuticals, Inc
roywbrown@sbcglobal.net

Daniel Decman
Lawrence Livermore National Laboratory
decman1@llnl.gov

Jack Faught
Spectra Gas Inc.
jackf@spectragasses.com

Donald F. Geesaman Ph.D. (co-chair)
Argonne National Laboratory
geesaman@anl.gov

Kenny Jordan
Association of Energy Service Companies
kjordan@aesc.net

Thomas H. Jourdan Ph.D.
University of Central Oklahoma
tjourdan@uco.edu

Steven M. Larson M.D.
Memorial Sloan-Kettering Cancer Center
larsons@mskcc.org

Richard G. Milner Ph.D.
Massachusetts Institute of Technology
milner@mit.edu

Jeffrey P. Norenberg Pharm.D.
University of New Mexico
jpnoren@unm.edu

Eugene J. Peterson Ph.D.
Los Alamos National Laboratory
ejp@lanl.gov

Lee L. Riedinger Ph.D.
University of Tennessee
lrieding@utk.edu

Thomas J. Ruth Ph.D.
TRIUMF
truth@triumf.ca

Susan Seestrom Ph.D. (ex-officio)
Los Alamos National Laboratory
seestrom@lanl.gov

Robert Tribble Ph.D. (ex-officio)
Texas A&M University
tribble@comp.tamu.edu

Roberto M. Uribe Ph.D.
Kent State University
ruribe@kent.edu

APPENDIX 3: AGENDAS OF MEETINGS I-V OF NSACI

NSAC Isotopes Subcommittee Meeting I
November 13-14, 2008
Hilton, Gaithersburg, Maryland

Thursday, November 13, 2008

9:00	Welcome
9:15	Charge from NSAC Chair – Robert Tribble
9:30	DOE-ONP perspective – Jehanne Simon-Gillo
10:00	Introduction – Don Geesaman
10:30	Break
10:45	Overview of the NE Isotopes Program – John Pantaleo
12:00	Lunch
1:30	Report from Isotopes Workshop – John D'Auria
2:15	Discussion of the charge and subcommittee perspective
3:30	Break
3:45	Industry perspective – Roy Brown
4:45	Discussion of the plan forward
5:30	Adjourn

Friday, November 14, 2008

9:00	Discussion of how to involve the broad community
10:00	Presentations of recent reports – Tom Ruth: National Academies Study
	Robert Atcher: National Cancer Institute Study
11:30	Executive session
1:00	Adjourn

NSAC Isotopes Subcommittee Meeting II
December 15-16, 2008
Bethesda, Maryland

Monday, December 15, 2008

9:00	Introduction
9:45	OMB – Mike Holland
10:00	FBI – Dean Fetteroff
10:45	Break
11:00	National Institute of Biomedical Imaging and Bioengineering –
	Belinda Seto
12:00	Lunch
1:30	Department of Homeland Security/DNDO – Jason Shergur
2:10	DOE Office of Nuclear Physics – John D'Auria
2:50	DOE Office of Basic Energy Sciences – Lester Morss
3:30	Break
3:45	National Science Foundation – Brad Keister

Isotopes for the Nation's Future: A Long Range Plan

4:30   Perspective – Jack Faught
5:00   Perspective – Kenny Jordan
5:30   Adjourn

Tuesday, December 16, 2008
9:00   National Cancer Institute – Craig Reynolds
9:40   NNSA – Victor Gavron
10:30   GNEP – Tony Hill
11:10   Executive session
1:30   Adjourn

NSAC Isotopes Subcommittee Meeting III
January 13-15, 2009
Rockville, Maryland

Tuesday, January 13, 2009
Input from Professional Societies and other groups on priorities for research.
9:00   Introduction
9:15   Sean O'Kelly, TRTR
10:00   Lynne Fairobent, AAPM
10:40   Break
11:00   Mark Stoyer, ACM/DNCT
11:40   J. David Robertson, MURR
12:30   Lunch
14:00   Gene Peterson, R&D for Accelerator Production of Isotopes
14:40   Scott Aaron, Stable Isotopes
15:30   Break
16:10   Roberto Uribe-Rendon, CIRMS
16:50   Robert Atcher, SNM
Wednesday, January 14, 2009
9:00   Michael Welch
9:40   Richard Toohey, HPS
10:30   Break
11:10-17:00 Executive Session

Thursday, January 15, 2009
9:00-16:00 Executive Session

NSAC Isotopes Subcommittee Meeting IV
February 13-15, 2009
Rockville, Maryland

Tuesday, February 10, 2009
Input from institutions and industry about present capabilities and future plans for isotope production.
9:00   Welcome

9:15	John Pantaleo, DOE NIPA
10:10	David Robertson, MURR
10:50	Break
11:10	Glen Young, ORNL
11:50	Jeff Binder, ORNL
12:30	Lunch
14:00	Leonard Mausner, BNL
14:40	Brad Sherrill, NSCL/FRIB
15:20	Richard Kouzes, PNNL
16:00	Break
16:15	Steve Laflin, International Isotopes
16:55	Ian Horn, NuView
17:3 5	Hugh Evans, Nuclitec

Wednesday, February 11, 2009

8:30	Doug Wells, Idaho State University
9:00	Donna Smith, LANL
9:40	Tracy Rudisill, SRNL
10:30	Richard Coats, SNL
11:10	Jim Harvey, Northstar
11:50	Frances Marshall, INL
12:30	Jerry Nolen, ANL
13:10	Executive Session

NSAC Isotopes Subcommittee Meeting V
March 25-27, 2009
Bethesda, Maryland

March 25, 2009
Open Session

9:00	Introduction, Geesaman
10:30	DOE-ONP Perspective, Gillo
11:00	FY09 Isotope Program Budget, Panteleo
11:30	BAC project, Brown

Closed Sessions

13:00	General Issues, Geesaman
13:30	Accelerator Options, Peterson
16:00	Reactor Options, Beierschmitt

March 26, 2009

8:00	Stable Isotope Options, Bier
10:00	R&D Required, Ruth
11:00	Program Operations, Riedinger
13:00	Workforce Development, Aprahamian
14:00	Discussion of Recommendations, Geesaman
16:00	Budgets, Geesaman

March 27, 2009
9:00   Review of Recommendations and Budgets
12:00   Plans to Complete Report

APPENDIX 4: LIST OF FEDERAL AGENCIES CONTACTED BY NSACI

Air Force Office of Scientific Research
Armed Forces Radiobiology Research Institute Department of Agriculture
Department of Defense
Department of Energy - Fusion Energy Sciences
Department of Energy - National Nuclear Security Administration - Nuclear Non-proliferation
Department of Energy - Basic Energy Sciences
Department of Energy - Biological and Environmental Research
Department of Energy - Nuclear Physics
Department of Homeland Security
Environmental Protection Agency
Federal Bureau of Investigation
National Cancer Institute
National Institute of Allergy and Infectious Disease
National Institute of Biomedical Imaging and Bioengineering
National Institute of Drug Abuse
National Institute of Environmental Health Science
National Institute of General Medical Science
National Institute of Standards and Technology
National Science Foundation - Directorate for Engineering
National Science Foundation - Directorate for Mathematical and Physical Sciences
National Science Foundation - Directorate for Biological Sciences
Office of Naval Research
State Department
U.S. Geologic Survey

APPENDIX 5: LIST OF PROFESSIONAL SOCIETIES CONTACTED BY NSACI

Academy of Molecular Imaging
Academy of Radiology Imaging
Academy of Radiology Research
American Association of Physicists in Medicine
American Association of Cancer Research
American Chemical Society
American Chemical Society - Division of Nuclear Chemistry and Technology
American College of Nuclear Physicians

American College of Radiology
American Medical Association American Nuclear Society
American Nuclear Society - Division of Isotopes and Radiation
American Pharmacists Association - Academy of Pharmaceutical Research and Science (APhAAPRS)
American Physical Society
American Physical Society - Division of Biological Physics
American Physical Society - Division of Material Physics
American Physical Society - Division of Nuclear Physics
American Society of Clinical Oncology
American Society of Hematology
American Society of Nuclear Cardiology
American Society of Therapeutic Radiation and Oncology
Council on Ionizing Radiation and Standards
Health Physics Society
National Organization of Test, Research and Training Reactors
Radiation Research Society
Radiation Therapy Oncology Group
Radiochemistry Society
Radiological Society of North America
Society of Molecular Imaging
Society of Nuclear Medicine

APPENDIX 6: LIST OF INDUSTRY TRADE GROUPS CONTACTED BY NSACI

Association of Energy Service Companies
Council on Radionuclides and Radiopharmaceuticals
Gamma Industry Processing Alliance
International Source Suppliers and Producers Association
Nuclear Energy Institute

APPENDIX 7: SUMMARY OF THE RESEARCH PRIORITIES AND RECOMMENDATIONS OF THE FIRST REPORT

Tables 8 and 9 of research priorities and the final recommendations of the first report of the NSACI subcommittee are copied here.

Isotopes for the Nation's Future: A Long Range Plan

Table 8. Research opportunities in medicine, pharmaceuticals and biology in order of relative priority

Research Activity	Isotope	Issue/Action
Alpha therapy	^{225}Ac ^{211}At ^{212}Pb	Current sources are limited. One valuable source for ^{225}Ac, extraction of ^{229}Th from ^{233}U may soon be lost.
Diagnostic dosimetry for proven therapeutic agents	^{64}Cu 86$_Y$ 124$_I$ ^{203}Pb	Used in conjunction with ^{67}Cu therapy 90Y therapy ^{131}I therapy and immune-diagnosis ^{212}Pb therapy The issue is the need for a coordinated network of production facilities to provide broad availability. There is need for R&D for common target and chemical extraction procedures.
Diagnostic tracer	^{89}Zr	Immune-diagnosis 3.27 d half-life allows longer temporal window for imaging of MoAbs, metabolism, bioincorporation, stemcell trafficking, etc.
Therapeutic	^{67}Cu	Requires specialized high-energy production facilities and enriched targets.

Table 9. Research opportunities in physical science and engineering in order of relative priority

Research Activity	Isotope	Issue/Action
Begin new facility to produce and study radioactive beams of nuclei from ^{252}Cf fission, for research in nuclear physics and astrophysics - CARIBU at ANL	^{252}Cf (2.6 yr)	Supply of ^{252}Cf is uncertain; 1 Ci source is needed each 1 1/2 year for at least four years.
Measure permanent atomic electric dipole moment of ^{225}Ra to search for time reversal violation, proposed to be enhanced due to effects of nuclear	^{225}Ra (15 d)	Supply of ^{225}Ra is limited. Need 10 mCi source of ^{225}Ra every two months for at least two years.
Create and understand the heaviest elements possible, all very short-lived and fragile. Study the atomic physics and chemistry of heavy elements for basic research and advanced reactor concepts	^{209}Po, ^{229}Th, ^{232}Th, ^{231}Pa, ^{232}U, ^{237}Np, ^{248}Cm, ^{247}Bk	Make certain actinides in HFIR and then prepare targets for accelerator-based experiments to make superheavy elements; targets needed are ^{241}Am, ^{249}Bk, ^{254}Es - not available now; need 10-100 mg on a regular basis; purity is important.

Table 9. (Continued)

Research Activity	Isotope	Issue/Action
Neutron detectors, electric dipole moment measurement, low temperature physics	^{3}He	Total demand exceeds that available.
Isotope dilution mass spectrometers	^{236}Np, ^{236,244}Pu, ^{243}Am, ^{229}Th	High purity ^{236}Np is not available; others are in limited supply; 10-100 mg needed on a regular basis; purity is important.
Search for double beta decay without neutrino emission - an experiment of great importance for fundamental symmetries	^{76}Ge	Need to fabricate large detectors of highly enriched ^{76}Ge; U.S. cannot produce quantity needed, ~1000 kg.
Spikes for mass spectrometers	^{202,203,205}Pb, ^{206}Bi, ^{210}Po	202,205Pb difficult to get in high purity in gram quantities.
Avogadro project - worldwide weight standard based on pure 28Si crystal balls	^{28}Si	Concern about future supply and cost of kg of material needed.
Radioisotope micro-power source	^{147}Pm, ^{244}Cm	Development needed for efficient conversion.
Isotopes for Mössbauer Spectroscopy, over 100 radioactive parent/stable daughter isotopes	^{57}Co, ^{119}mSn ^{67}Ni, ^{161}Dy, ...	Some Isotopes only available from Russia, a concern for scientific community.

FIRST REPORT: RECOMMENDATIONS FOR CHARGE 1

Compelling research opportunities were identified and presented in prioritized lists within the two areas of 1) biology, medicine, and pharmaceuticals, and 2) physical sciences and engineering. The third area 3) security applications did not have immediate research priorities but made a number of observations and recommendations that apply more broadly for the entire NIPA program. While it is challenging to assess relative scientific merit across disciplines, we have identified the highest priorities for the most compelling research opportunities. These recommendations also define the relative priorities of opportunities in Tables 8 and 9.

There are compelling research opportunities using alpha-emitters in medicine. There is tremendous potential in developing far more effective treatments of cancers by the use of alpha- emitters in comparison to other radioisotopes. Therefore, development and testing of therapies using alpha emitters are our highest priority for research isotope production for the medical field. This priority is reinforced by the potential need for rapid action due to the 2012 deadline for downblending of current DOE stocks of ^{233}U, a procedure that would eliminate its value as a source of ^{225}Ac.

1. Invest in new production approaches of alpha-emitters with highest priority for ^{225}Ac. Extraction of the thorium parent from ^{233}U is an interim solution that needs to be seriously considered for the short term until other production capacity can become available.

There is strong evidence for the potential efficacy of pairs of isotopes with simultaneous diagnostic/therapeutic capabilities. Table 8 of this report presents a prioritized list isotopes that have the greatest research potential in Biology, Medicine, and Pharmaceuticals. NSACI finds the research opportunities offered with these pairs of isotopes to be the second highest priority in identifying compelling research opportunities with isotopes. Many of these isotopes could be produced at existing accelerator facilities. We recommend the maximization of the production and availability of these isotopes domestically in the U.S. through investments in research and coordination between existing accelerators. The panel felt that such a network could benefit all areas of basic research and applications from security to industry. This should include R&D to standardize efficient production target technology and chemistry procedures.

2. We recommend investment in coordination of production capabilities and supporting research to facilitate networking among existing accelerators.

The basic physical sciences and engineering group prioritized research opportunities across various disciplines and a summary of this prioritization is given in Table 9. The availability of californium, radium, and other transuranic isotopes, the first three opportunities in Table 9, are particularly important for research.

3. We recommend the creation of a plan and investment in production to meet these research needs for heavy elements.

Experts in the nuclear security and applications areas strongly consider the vulnerability of supply from foreign sources to be of highest priority. This concern was echoed strongly by all members of the subcommittee in from medicine to basic science and engineering. Additionally, the projected demand for ^{3}He by national security agencies far outstrips the supply. This would likely endanger supply for many other areas of basic research. While it is beyond our charge, it would be prudent for DOE/NNSA and DHS to seriously consider alternative materials or technologies for their neutron detectors to prepare if substantial increases in ^{3}He production capacity cannot be realized.

4. We recommend a focused study and R&D to address new or increased production of ^{3}He.

The remaining isotopes in Tables 8 and 9 all are promising research opportunities, and funds for production from the Research Isotope Development and Production Subprogram would be well spent on targeted production of these isotopes to meet immediate research needs, especially if unique production opportunities arise. However, at this point in prioritization, NSACI concludes that larger, long-term issues should take priority. The darker tone of blue used in Table 9 is an indication of that.

An important issue for the use of isotopes is the availability of high-purity, mass-separated isotopes. The stable isotopes ^{76}Ge and ^{28}Si (^{3}He is stable but obtained from the beta-decay of ^{3}H, not by isotope separation) listed in Table 9 are needed in large quantities that present special problems. While no other individual stable isotope reached the level of the highest research priority, the broad needs for a wide range of mass-separated isotopes and the prospect of no domestic supply raised this issue in priority for the subcommittee. NSACI feels that the unavailability of a domestic supply poses a danger to the health of the national research program and to national security. NSACI recommends:

5. Research and Development efforts should be conducted to prepare for the reestablishment of a domestic source of mass-separated stable and radioactive research isotopes.

Vital to the success of all scientific endeavors is the availability of trained workforce. While the scientific opportunities have expanded far beyond the disciplines of radiochemistry and nuclear chemistry, the availability of trained personnel remains critical to the success of research in all frontiers of basic science, homeland security, medicine, and industry. The individual research areas must make concerted efforts to invest in work-force development to meet these needs. The isotope program has a special responsibility to ensure a trained workforce in the production, purification and distribution of isotopes.

6. We recommend that a robust investment be made into the education and training of personnel with expertise to develop new methods in the production, purification and distribution of stable and radioactive isotopes.

All of the issues and recommendations considered here will be important input for answering the 2nd NSACI charge (See Appendix 1) due in July 31, 2009, developing a long range plan for the Nuclear Isotopes Production and Application Program.

INDEX

#

2009 appropriations bill, vii, 2, 4
21st century, vii, 1, 5, 131

A

accelerator, 3, 13, 15, 16, 28, 32, 47, 72, 86, 92, 93, 94, 95, 96, 97, 98, 99, 100, 101, 102, 108, 112, 115, 119, 120, 121, 122, 124, 126, 131, 133, 136, 137, 144, 145, 151, 154, 156, 158, 171, 173
access, 24, 46, 47, 60, 102, 112, 120, 148, 153, 157, 158
accreditation, 125
acid, 16, 91, 93, 94, 146
activity level, 150
acute myeloid leukemia, 52
adults, 21, 57
advisory body, 149
age, 10, 129, 144
agencies, 3, 5, 6, 33, 35, 45, 69, 72, 74, 85, 101, 112, 120, 124, 130, 133, 149, 152, 156, 162, 173
agility, 46, 101, 112, 153, 157
agriculture, vii, 2, 5
Air Force, 40, 169
alternative energy, 73
amalgam, 67
American Recovery and Reinvestment Act, 47, 54, 139, 141, 144, 153, 158
americium, 61
amino, 21, 58, 81
amino acid, 21, 58, 81
ammonia, 26, 64
angioplasty, 106
antibody, 19, 55, 142
antimatter, 25, 62
antiparticle, 25, 62

appropriations, vii, 2, 4, 74, 134, 138, 141, 147, 153, 155
Appropriations Act, 51, 52
argon, 8, 25, 63, 84
arthritis, 18, 55, 106
assessment, 49, 82, 121
assets, 107, 111, 112
atomic nucleus, 51, 60, 102
atoms, 18, 25, 27, 30, 51, 54, 61, 62, 64, 65, 66, 71, 120
Atoms for Peace, 5
attribution, 125, 128, 135
Austria, 111
authentication, 70
avoidance, 98

B

BAC, 168
banks, 16, 92
base, vii, 1, 45, 47, 84, 85, 101, 102, 107, 112, 124, 152, 153, 155, 156, 158
basic research, 3, 24, 28, 32, 33, 35, 60, 140, 162, 171, 173
batteries, 28
beams, 24, 25, 28, 52, 60, 61, 63, 78, 89, 90, 98, 102, 112, 145, 148, 171
behaviors, 25, 63
Belgium, 27, 65
benefits, 74, 97, 127, 132, 148
berkelium, 61
beta particles, 19, 20, 56, 57
bioavailability, 34
biochemistry, 8, 77, 84
bio-fuels, vii, 2, 5
biological processes, 78
biological sciences, 5, 74
biological systems, 21, 34, 58, 78

biomolecules, 19, 20, 55, 56
blood, 18, 55, 125
blood flow, 125
bonding, 26, 64
bone, 16, 18, 55, 94, 106
bone cancer, 106
bone marrow, 18, 55
brachytherapy, 78
brain, 56, 106
brain tumor, 56, 106
breeding, 24
Brillouin light scattering, 27, 64
building blocks, 21, 58, 81
burnout, 116
business model, 144
businesses, 86
buyers, 148

C

calibration, 16, 26, 29, 63, 69, 70, 71, 75, 78, 94, 116, 145
californium, 3, 32, 61, 103, 173
calorimetry, 69
cancer, 16, 18, 19, 20, 50, 55, 56, 57, 75, 94, 104, 105, 106, 134, 142, 145
cancer cells, 20, 56
candidates, 71, 123
capsule, 112, 151, 152
carbon, 7, 23, 26, 51, 60, 63, 77, 78, 83, 102, 105
carcinoid tumor, 19, 55
casting, 85
category a, 45, 156
cell line, 18, 54
ceramic, 85, 141
certification, 46, 125, 150, 151, 153, 157
challenges, 6, 36, 48, 72, 73, 75, 99, 129, 139, 144, 163
chelates, 18, 55
chemical, 1, 7, 16, 18, 20, 21, 23, 26, 43, 51, 53, 55, 56, 58, 61, 64, 67, 83, 85, 89, 92, 93, 97, 99, 114, 119, 121, 122, 126, 127, 131, 140, 171
chemical characteristics, 121
chemical properties, 67
chemical reactions, 21, 58
Chicago, 127
childhood, 56
children, 21, 57
chromatography, 18, 54
cleaning, 146
cleanup, 85
climate, 16, 26, 63, 94
climate change, 26, 63

climates, 26, 64
clinical application, 19, 55, 103
clinical trials, 11, 17, 18, 54, 55, 56, 58, 73, 142, 147
closure, 9, 72, 151
coal, 75
coherence, 27, 64
collaboration, 25, 27, 53, 62, 63, 65, 94, 95, 108, 111, 130, 134, 149
College Station, 34, 161
colleges, 9
commercial, 7, 11, 13, 14, 15, 17, 21, 22, 24, 35, 36, 44, 45, 46, 48, 50, 51, 52, 54, 58, 59, 67, 73, 74, 75, 76, 78, 81, 82, 85, 86, 89, 95, 97, 98, 100, 101, 102, 103, 106, 112, 118, 122, 133, 134, 135, 143, 147, 148, 149, 150, 151, 152, 153, 154, 156, 157, 158, 162, 163
communication, 46, 149, 150, 153, 157
communities, 2, 6, 35, 44, 49, 80, 106, 148, 150, 162
community, 4, 10, 15, 22, 27, 28, 35, 39, 51, 60, 65, 74, 81, 82, 89, 90, 107, 110, 111, 120, 121, 124, 133, 146, 147, 148, 149, 162, 166, 172
compatibility, 99
compensation, 50
competition, 49, 50, 133
competitiveness, 140, 153
complement, 13, 86, 93, 99, 100
compliance, 89, 146
complications, 106
composition, 27, 65
compounds, 7, 15, 18, 20, 26, 56, 63, 64, 83, 144
computation, 128
computer, 68, 130
condensation, 25, 63
conductivity, 27, 64
configuration, 85, 112
congress, 76, 129, 142
consensus, 81, 99, 112
conservation, 25, 62
constituents, 78
construction, 9, 16, 36, 47, 48, 63, 72, 92, 93, 111, 112, 122, 125, 153, 154, 155, 158, 163
containers, 94, 108, 112, 149, 150, 151
contaminant, 21, 58
contamination, 76, 80, 114, 115
cooling, 26, 63, 80, 91, 98, 112
cooperation, 25, 63, 97, 100, 121, 126
coordination, 3, 22, 32, 59, 99, 101, 107, 108, 121, 149, 150, 152, 173
copper, 25, 63
cost, 7, 8, 16, 18, 28, 35, 47, 51, 52, 54, 72, 74, 75, 76, 80, 83, 89, 90, 92, 95, 96, 98, 102, 107, 115, 117, 133, 140, 142, 144, 146, 147, 148, 150, 154, 155, 158, 162, 172

cost benefits, 98
cost saving, 98
covering, 104
CPT, 25
CRP, 13, 14
crystal structure, 66
crystalline, 27, 65
crystals, 25, 27, 62, 64
cure, 18, 20, 55, 56
curium, 61
curriculum, 46, 102, 112, 124, 131, 132, 157
curriculum development, 46, 102, 112, 124, 132, 157
customers, 11, 45, 46, 53, 72, 85, 91, 92, 94, 101, 112, 115, 134, 137, 147, 148, 150, 151, 152, 153, 156, 157
cycles, 78, 104, 106
cycling, 26, 64
cytotoxicity, 18, 54

D

danger, 3, 33, 174
data center, 140, 141
decay, 3, 7, 13, 19, 20, 23, 25, 27, 28, 33, 34, 51, 53, 57, 60, 62, 64, 66, 70, 80, 81, 85, 86, 87, 88, 110, 114, 115, 116, 117, 119, 122, 142, 172, 173
deficiencies, 96
deformation, 25, 28, 61, 171
Department of Agriculture, 40, 169
Department of Defense, 40, 69, 169
Department of Energy, vii, 2, 4, 5, 9, 15, 34, 40, 43, 47, 48, 50, 56, 60, 72, 75, 86, 103, 131, 133, 153, 161, 169
Department of Homeland Security (DHS), 3, 29, 30, 31, 33, 39, 40, 69, 118, 166, 169, 173
Department of Transportation, 151
depreciation, 51
destruction, 117
detectable, 30
detection, 8, 23, 29, 69, 71, 75, 76, 78, 84, 115, 116, 125
detection system, 29, 69, 125
deuteron, 13, 86
diffusion, 7, 8, 82, 83, 84
diodes, 25, 63
dipole moments, 61
disclosure, 70
diseases, 104, 105, 127, 134
dismantlement, 53, 67, 117
displacement, 24, 60
disposition, 30, 51, 70
distillation, 1, 7, 43, 83
distilled water, 146

distribution, 4, 11, 15, 19, 33, 57, 85, 89, 94, 110, 116, 117, 135, 138, 139, 141, 174
diversity, 149
DMF, 15
DNA, 104
dogs, 78
draft, 50
drawing, 128
drug discovery, vii, 2, 5
drugs, 18, 55
dwindling population, 46, 157

E

ecology, 60
economies of scale, 85
ecosystem, 26, 64
education, 2, 4, 6, 9, 23, 33, 75, 96, 104, 124, 125, 129, 132, 146, 149, 160, 174
effluents, 93
effort level, 47, 153, 155, 158
electric field, 24, 61
electromagnetic, 1, 8, 43, 47, 51, 80, 84, 85, 86, 93, 123, 154, 158
electron, 13, 19, 25, 27, 34, 55, 57, 64, 66, 86, 99, 116, 120, 122
electron microscopy, 116
electrons, 19, 51, 56, 57
elementary particle, 24, 60, 62, 80
emergency, 50, 71
emission, 13, 19, 28, 50, 66, 70, 71, 75, 86, 91, 110, 116, 122, 146, 151, 172
emitters, 2, 20, 32, 56, 122, 124, 172, 173
encapsulation, 98
endocrine, 19, 55
energy, vii, 2, 5, 13, 14, 15, 16, 17, 19, 20, 22, 23, 27, 29, 31, 47, 56, 57, 59, 60, 61, 64, 65, 70, 71, 74, 75, 81, 86, 87, 89, 90, 92, 93, 94, 95, 96, 97, 98, 99, 100, 101, 102, 110, 112, 115, 119, 120, 121, 122, 129, 131, 132, 145, 146, 158, 171
energy density, 29, 65
energy efficiency, 29, 65
energy expenditure, 81
energy transfer, 19
engineering, vii, 1, 2, 3, 7, 9, 23, 29, 30, 32, 33, 60, 63, 67, 69, 71, 121, 125, 128, 130, 132, 151, 171, 172, 173
environment, 21, 26, 30, 44, 49, 58, 60, 61, 64, 71, 72, 80, 114, 118, 119
environmental conditions, 68
environmental issues, 80
Environmental Protection Agency (EPA), 5, 40, 146, 169

equipment, 10, 51, 70, 75, 91, 108, 111, 128, 141, 142

Europe, 11

everyday life, 36, 163

evidence, 2, 26, 32, 64, 173

evolution, 30, 71, 89, 133, 154

exaggeration, 21, 58, 81

excess supply, 118

excretion, 102

execution, 73

exercise, 101

expenditures, 147

expertise, 4, 6, 29, 33, 53, 69, 72, 85, 107, 112, 124, 129, 131, 174

explosives, 75

export control, 9, 85, 97

exposure, 2, 30, 71, 131, 151

extraction, 23, 34, 73, 115, 171

F

fabrication, 16, 68, 85, 90, 96, 107, 109, 112, 140, 141, 145, 147

fairness, 50

FDA, 14, 15, 19, 20, 22, 55, 56, 59, 91, 92, 97, 142

FDA approval, 97

Federal Bureau of Investigation (FBI), 29, 39, 40, 69, 166, 169

federal facilities, 135

federal government, 5, 50, 73, 80, 132

Federal Register, 50

feedstock, 47, 85, 86, 116, 117, 158

fertilizers, 26, 64

filters, 91

filtration, 91

financial, 76, 99, 145

financial resources, 99

fission, 19, 24, 28, 61, 70, 75, 76, 102, 103, 112, 114, 117, 120, 122, 171

flexibility, 85, 98, 112

flights, 150

fluctuations, 6, 74, 100, 107, 130, 141, 144, 154

fluid, 29, 65, 67

foils, 85

food, 23, 60

food safety, 23, 60

force, 4, 6, 33, 75, 156, 174

formation, 24, 26, 27, 35, 60, 64, 162

foundations, 2

fragility, 122

fragments, 70

France, 115, 149

freedom, 23, 60

funding, 8, 9, 36, 45, 47, 48, 52, 74, 84, 90, 95, 101, 107, 126, 131, 133, 134, 138, 139, 140, 144, 145, 147, 153, 154, 155, 156, 158, 163

funds, 33, 47, 72, 111, 133, 137, 139, 140, 141, 144, 147, 150, 153, 154, 155, 158, 173

fusion, 67, 78, 80, 134

G

Galileo, 28, 65, 68

gamma rays, 70

geology, 23, 60

geometry, 122

germanium, 25, 27, 63, 64

Germany, 27, 65

global scale, 78

governance, 96

governments, 73, 74, 125

graduate education, 131

graduate students, 131, 132

grants, 10, 120

groundwater, 146

growth, 19, 55, 126, 129, 151

growth hormone, 19, 55

guidance, 22, 59, 153

guidelines, 71

H

half-life, 14, 56, 59, 65, 75, 106, 116, 119, 171

halogens, 82

health, vii, 1, 3, 5, 8, 33, 45, 50, 72, 74, 76, 80, 84, 129, 158, 174

health care, vii, 2, 5, 8, 45, 76, 84, 158

health care system, 45, 76, 158

health enterprises, vii, 1

heat capacity, 26, 63

height, 91

hemisphere, 106

high school, 132

hiring, 131

history, 26, 53, 63, 77, 97, 119

HIV, 142

homeland security, vii, 1, 2, 4, 5, 6, 33, 44, 49, 124, 125, 129, 132, 134, 174

host, 52, 112, 138, 149

human, 18, 21, 50, 54, 55, 57, 60, 67, 71, 72, 81, 104, 105, 145

human body, 60

human health, 67

hydrogen, 26, 63, 77, 100

hyperfine interaction, 27, 64

Index

I

ideal, 11, 27, 64, 75, 116
identification, 4, 35, 44, 49, 162
identity, 99
image, 21, 58
immune system, 18, 54
immunotherapy, 19
implants, 16, 94
improvements, 29, 31, 65, 72, 99, 108, 140, 147, 150
impurities, 27, 65, 98
in transition, 27, 64
in vitro, 17, 54
in vivo, 17, 18, 21, 50, 54, 57, 78, 102
income, 9
individuals, 49, 76, 100, 107, 124, 125, 154, 155
industry, 2, 3, 4, 5, 6, 7, 9, 11, 15, 21, 32, 33, 35, 44, 49, 50, 56, 58, 74, 75, 77, 81, 90, 98, 104, 114, 115, 124, 126, 127, 129, 130, 133, 140, 162, 167, 173, 174
inflation, 137, 138
information sharing, 149
information technology, 149
infrastructure, 5, 9, 36, 44, 47, 48, 53, 96, 100, 101, 107, 109, 112, 130, 133, 137, 138, 140, 141, 142, 144, 146, 147, 152, 153, 154, 155, 156, 158, 163
initiation, 75
injections, 142
insertion, 91
institutions, 44, 49, 120, 131, 142, 167
intellectual capital, 76
intellectual property, 74
intellectual property rights, 74
intelligence, 125, 128
International Atomic Energy Agency, 111
investment, 3, 4, 29, 32, 33, 80, 114, 128, 137, 140, 142, 145, 147, 150, 152, 173, 174
investments, 3, 32, 36, 45, 47, 48, 72, 87, 99, 118, 138, 142, 152, 153, 155, 156, 158, 163, 173
iodine, 18, 50, 55, 102
ionization, 78
ionizing radiation, 71
ions, 51, 90, 100
iron, 34, 78
irradiation, 16, 17, 22, 31, 34, 59, 70, 72, 73, 91, 92, 94, 96, 98, 104, 106, 110, 112, 114, 115, 116, 117, 119, 142, 143, 145
isolation, 124
isomers, 51
Isotope Development and Production for Research and Applications Program (IDPRA), viii, 44

issues, vii, 1, 2, 4, 5, 14, 20, 33, 45, 46, 49, 50, 52, 56, 68, 73, 74, 75, 76, 80, 85, 89, 97, 107, 116, 148, 149, 150, 152, 153, 157, 158, 173, 174
Italy, 25, 27, 62, 65

J

Japan, 27, 65
joint ventures, 13
Jordan, 37, 39, 164, 167
justification, 148

K

kill, 19, 20, 55, 57
krypton, 7, 8, 83, 84

L

labeling, 18, 27, 54, 64, 77
lakes, 64
landscape, 6, 7, 49, 135
law enforcement, 125
lead, 23, 25, 27, 45, 53, 60, 62, 76, 80, 91, 107, 114, 122, 124, 141, 144, 158
leadership, 46, 157
lending, 60
lepton, 25, 62
leukemia, 18, 55, 115
LIFE, 103
life cycle, 87, 98
light, 51, 62, 67, 70, 77, 81, 109, 112, 117
limestone, 26, 64
liquid chromatography, 91
liquids, 23, 60, 150
lithium, 67, 80
liver, 56
liver cancer, 56
localization, 20
logging, 12, 50, 115
logistics, 89, 94, 97, 149
low temperatures, 25, 63
lower prices, 52
lymphoma, 19, 55

M

magnetic field, 51
magnetic resonance, 30, 78
magnetic resonance imaging, 30, 78
magnetism, 27, 60, 64

magnets, 146
magnitude, 22, 27, 59, 62, 64, 99, 102, 110
majority, 8, 9, 17, 18, 54, 55, 84, 103
man, 21, 51, 58, 78, 81, 105
management, 6, 46, 48, 102, 105, 140, 149, 153, 157
manpower, 111, 154, 155, 156
mapping, 125, 127
marginal costs, 73
marketing, 16, 90
marrow, 18, 55
Mars, 68
Maryland, 14, 38, 39, 127, 166, 167, 168
mass, 3, 20, 23, 24, 27, 28, 29, 33, 51, 60, 65, 69, 70,
    78, 116, 140, 146, 172, 173, 174
mass spectrometry, 3, 78
material sciences, vii, 1, 5
materials, 3, 11, 16, 17, 22, 23, 26, 29, 31, 33, 44,
    50, 51, 52, 59, 60, 64, 65, 66, 68, 69, 70, 71, 72,
    74, 78, 85, 92, 94, 98, 99, 103, 104, 106, 112,
    119, 122, 123, 125, 126, 128, 133, 135, 142, 146,
    173
materials science, 78, 99, 128
matter, 25, 27, 61, 62, 63, 64, 120
measurement, 25, 27, 28, 62, 63, 64, 70, 116, 172
measurements, 20, 25, 26, 29, 31, 56, 63, 64, 69, 70,
    72, 81, 100, 121
median, 19
medical, 2, 5, 11, 15, 17, 22, 29, 30, 32, 44, 45, 46,
    47, 50, 51, 54, 59, 62, 65, 72, 76, 77, 78, 102,
    104, 105, 106, 120, 124, 125, 127, 130, 132, 133,
    134, 150, 157, 158, 172
medicine, vii, 1, 2, 3, 4, 5, 7, 11, 15, 18, 19, 32, 33,
    35, 50, 52, 55, 56, 74, 76, 86, 89, 90, 105, 122,
    124, 125, 127, 129, 130, 131, 132, 162, 171, 172,
    173, 174
membership, 2, 6, 44, 49
memory, 27, 64
mercury, 67, 80, 83
messages, 154
metabolism, 21, 23, 34, 57, 77, 78, 81, 102, 171
metals, 85
metastasis, 106
meteorites, 27, 64
meter, 150
methodology, 99
Mexico, 38, 165
microorganisms, 26, 64
migration, 61
milligrams, 81
mission, 11, 17, 35, 44, 46, 47, 49, 52, 56, 72, 76, 92,
    96, 97, 102, 104, 106, 107, 116, 131, 132, 133,
    139, 149, 157, 158, 162
missions, 6, 9, 11, 47, 52, 75, 89, 96, 98, 103, 157

Missouri, 10, 11, 53, 93, 104, 122, 134
modernization, 144
molecules, 26, 64, 77
molybdenum, 50
Mössbauer effect, 66
MTS, 93
multiplication, 70
myelodysplasia, 18, 55
myocardial infarction, 143

N

nanostructures, 27, 64
National Aeronautics and Space Administration, 106
National Institutes of Health, 5
national interests, 50
National Isotope Production and Application
    program (NIPA), vii, 2
National Research Council, 22, 49, 58, 131, 160
national security, 2, 3, 5, 7, 30, 33, 35, 44, 63, 74, 75,
    77, 129, 162, 173, 174
national strategy, 36, 48, 50, 163
natural resources, 23, 60
neon, 8, 84
neptunium, 61
networking, 3, 22, 32, 45, 59, 101, 112, 152, 156,
    173
neuroblastoma, 56
neutrons, 29, 30, 31, 51, 60, 61, 69, 70, 71, 72, 74,
    75, 94, 102, 120, 122
New England, 56
next generation, 76, 108, 112, 132, 151
nitrite, 26, 64
nitrogen, 7, 21, 23, 26, 58, 60, 63, 64, 77, 78, 81, 83
nitrogen gas, 26, 64
nitrous oxide, 26, 64
Nobel Prize, 66
North America, 7, 41, 81, 159, 170
NPS, 13
NPT, 12
NSA, 5
NSAC Isotope Subcommittee (NSACI), vii, 2
nuclear fuels, vii, 2
nuclear medicine physician, 125, 127
Nuclear Science Advisory Committee (NSAC), vii,
    viii, 2, 4, 34, 44, 48, 161
nuclear weapons, 5, 53, 56, 67, 70, 74, 80, 117
nuclei, 24, 25, 28, 51, 61, 62, 66, 119, 145, 171
nucleus, 19, 25, 51, 57, 60, 61, 62
nuclides, 8, 21, 58, 78, 81, 83, 93, 110, 122
nutrient, 21, 26, 58, 64, 81
nutrients, 21, 26, 58, 64, 78
nutrition, vii, 2, 5, 8, 50, 51, 81, 84

Index

181

O

obesity, 159
Office of Management and Budget, 6
Office of Nuclear Energy (NE), vii, 5, 43, 47
Office of Nuclear Physics (ONP), vii, 44, 48
Office of Science in Nuclear Physics, vii, 2
oil, 50, 73, 106, 115
Oklahoma, 37, 164
operating costs, 90, 98, 115, 145, 146, 155
operations, 8, 11, 36, 48, 76, 80, 91, 93, 101, 106, 112, 140, 141, 144, 145, 147, 154, 155, 163
opportunities, vii, viii, 2, 3, 4, 6, 7, 21, 23, 24, 25, 29, 31, 32, 33, 35, 36, 44, 47, 48, 49, 58, 63, 69, 72, 93, 99, 102, 118, 124, 131, 132, 154, 158, 162, 163, 171, 172, 173, 174
optimization, 122
orbit, 68
organ, 76
organic solvents, 120
outreach, 132, 149
overlap, 6
oversight, 133
ownership, 14
oxidation, 26, 64
oxygen, 7, 26, 63, 77, 83

P

Pacific, 53, 56, 134, 135, 138
pain, 16, 94
palliative, 18, 55, 106
parallel, 20, 57, 94
parity, 24
particle bombardment, 13, 86
passenger airline, 150
PCA, 112
peace, 103
peer review, 74
peptide, 19, 55
peptides, 19, 20, 56
perfusion, 5, 97, 127
permeability, 102
permit, 91
PET scan, 16, 78, 81, 94, 143
petrochemical, vii, 2, 5
pharmaceutical, 5
pharmaceuticals, 2, 7, 32, 124, 171, 172
phosphate, 18, 55
phosphorus, 26, 64, 78, 102, 104
photons, 70, 71, 122
photovoltaic devices, 29, 65

physical sciences, 2, 3, 7, 29, 32, 69, 124, 132, 172, 173
physicians, 76
physics, vii, 1, 3, 5, 6, 23, 24, 25, 27, 28, 29, 30, 31, 60, 61, 63, 64, 69, 70, 71, 72, 78, 80, 90, 97, 99, 124, 125, 127, 128, 130, 131, 132, 145, 171, 172
planets, 60
plants, 70
platinum, 80, 83
plutonium, 30, 45, 53, 61, 69, 70, 133
policy, 8, 11, 46, 50, 52, 74, 147, 157
polycythemia, 18, 55
polycythemia vera, 18, 55
pools, 74, 112
population, 19
portfolio, 85, 93
positron, 13, 17, 50, 51, 78, 81, 86, 120
positron emission tomography (PET), 5, 13, 14, 15, 16, 50, 51, 78, 81, 86, 89, 94, 95, 99, 100, 119, 120, 121, 143, 145
positrons, 57
power generation, 103
predictability, 82
preparation, vii, 2, 4, 31, 34, 48, 72, 133, 161
president, vii, 2, 4, 5, 34, 35, 36, 47, 48, 52, 76, 137, 139, 153, 158, 161, 162, 163
prevention, 29, 65
principles, 24, 61, 82
private sector investment, 8, 84
privatization, 55, 56
probability, 102, 119
probe, 24, 61, 66, 78
producers, 7, 13, 22, 35, 50, 59, 76, 81, 86, 162
production costs, 146
production targets, 75, 106
production technology, 76
profit, 148
program staff, 147
project, 16, 27, 28, 65, 72, 76, 92, 133, 140, 141, 168, 172
proliferation, 19, 40, 45, 53, 55, 74, 76, 158, 169
prostate cancer, 16, 19, 55, 94, 106
protection, 125
proteins, 21, 58, 81
protons, 15, 16, 17, 51, 60, 74, 90, 92, 94, 97, 98, 145
prototype, 18, 55
public broadcasting, 132
public sector, 100
public-private partnerships, 155
purification, 4, 33, 174
purity, 3, 4, 21, 27, 28, 33, 35, 58, 65, 74, 80, 85, 115, 119, 122, 123, 146, 162, 171, 172, 173

Q

quality assurance, 96, 145
quality control, 16, 90
quantum field theory, 25
quantum mechanics, 24, 61

R

R&D investments, 138
radiation, 18, 19, 20, 29, 30, 41, 49, 55, 56, 57, 65,
66, 69, 71, 114, 125, 127, 150, 160, 170
radiation damage, 29, 65
radio, 2, 4, 19, 32, 33, 68, 116
radioactive, vii, viii, 1, 2, 3, 4, 5, 18, 20, 21, 22, 23,
24, 25, 28, 29, 30, 31, 33, 34, 35, 43, 44, 45, 47,
48, 50, 51, 54, 55, 56, 57, 59, 60, 61, 62, 66, 69,
70, 71, 72, 74, 76, 78, 85, 86, 91, 93, 101, 102,
119, 125, 126, 127, 133, 134, 142, 146, 147, 157,
158, 162, 171, 172, 174
radioactive isotopes, viii, 1, 2, 4, 5, 23, 29, 31, 35,
43, 44, 47, 48, 60, 69, 70, 71, 78, 85, 86, 102,
125, 127, 133, 134, 157, 158, 162, 174
radioactive tracer, 21, 57
radioactive waste, 91, 146
radiography, 75, 103
radioisotope, 15, 28, 65, 68, 70, 75, 86, 90, 94, 105,
106, 110, 111, 120, 142, 145, 147, 150, 151
radiotherapy, 19, 122
radium, 3, 32, 62, 173
raman spectroscopy, 27, 64
ramp, 146
reactions, 16, 24, 26, 30, 31, 50, 51, 61, 64, 70, 71,
72, 76, 87, 89, 92, 94, 98, 102, 121, 145
reading, 151
reality, 122
recommendations, 2, 4, 7, 22, 29, 32, 33, 35, 44, 45,
47, 49, 58, 76, 82, 99, 101, 124, 126, 132, 134,
156, 158, 162, 170, 172, 174
recovery, 8, 35, 51, 52, 80, 115, 117, 118, 133, 162
red blood cells, 18, 55, 102
regulatory requirements, 132
relaxation, 64
relevance, 131
reliability, 10, 30, 71, 82, 108, 122
remission, 142
repair, 91, 142
reprocessing, 67
requirements, 16, 21, 27, 29, 58, 63, 64, 65, 68, 76,
78, 92, 95, 98, 106, 144, 151
research facilities, 7
research funding, 120, 147

researchers, 11, 17, 22, 35, 46, 52, 59, 74, 104, 106,
110, 112, 120, 133, 148, 153, 156, 157, 162
resolution, 27, 65, 69, 91
resource development, vii, 2, 5
resources, 6, 11, 36, 45, 48, 74, 75, 76, 95, 100, 118,
120, 122, 131, 132, 137, 140, 147, 149, 154, 155,
156, 163
response, 4, 35, 45, 60, 111, 142, 146, 154, 156, 162
retirement, 126
retirement age, 126
revenue, 54, 91, 100, 134, 135, 136, 137, 138, 139,
141, 144, 147, 154
rights, 76
risk, 10, 47, 73, 82, 155, 158
risks, 74, 76
routes, 17, 119
rubidium, 97
rules, 133
Russia, 11, 13, 21, 27, 28, 31, 58, 65, 75, 81, 94, 97,
149, 172

S

safety, 10, 29, 30, 65, 71, 72, 93, 103, 110, 115
saturation, 119
scaling, 25, 63
scattering, 11, 25, 27, 52, 63, 64, 66, 67, 97, 104
school, 92, 125, 127, 132
science, vii, 1, 2, 3, 4, 6, 16, 23, 31, 33, 43, 60, 64,
72, 75, 92, 99, 101, 102, 112, 125, 128, 129, 130,
131, 132, 171, 173, 174
scientific discoveries, vii, 1, 5
scientific understanding, 30, 71
scope, 8, 17, 36, 48, 84, 85, 147, 163
security, 3, 8, 9, 24, 29, 30, 31, 32, 69, 70, 73, 74,
85, 108, 125, 151, 172, 173
seed, 16, 94
selectivity, 102
self-sufficiency, 149
sellers, 136
semiconductors, 27, 64
sensitivity, 25, 61, 62
services, 35, 44, 46, 51, 52, 56, 73, 85, 101, 133,
135, 138, 153, 157, 162
sewage, 26, 64
short supply, 5, 8, 83, 84, 140
shortage, 34, 69, 72, 81, 118, 129, 130, 132
shortfall, 30, 81, 129, 132
showing, 66, 90
signs, 50
silicon, 27, 64, 65
Sinai, 15
skilled personnel, 76, 126, 130

Index

183

skilled workers, 46, 157
SNS, 25, 63
sodium, 18, 55
solar system, 26, 64
solid state, 116
solution, 2, 32, 47, 75, 78, 101, 118, 122, 158, 173
South Africa, 94, 97
Soviet Union, 133
spare capacity, 17
specialists, 126, 130
species, 31, 72, 119, 120, 124
specifications, 21, 58, 94
spectroscopy, 27, 29, 64, 69, 91
spin, 24, 27, 60, 61, 64, 66
stability, 24, 29, 61, 69, 78, 110, 139
staffing, 144, 147
stakeholders, viii, 5, 35, 44, 48, 150, 162
Standard Model, 24, 25, 60, 61, 62
standing subcommittee, vii, 2, 4
stars, 51, 60
state, 19, 24, 26, 27, 51, 52, 55, 64, 66, 78, 81, 95, 104, 116
states, 18, 27, 51, 55, 64, 66, 67, 74
steel, 91
storage, 74, 91, 93, 116, 118, 145
stressors, 103
strontium, 56
structure, 14, 23, 24, 27, 60, 61, 64, 65, 76, 96
subscribers, 75
subsidy, 147
substitution, 27, 64
substrate, 26, 64
substrates, 78
sulfur, 77
superconductivity, 25, 27, 63, 64
superfluidity, 25, 63
superheavy elements, 28, 171
supplier, 11, 76, 104, 115
suppliers, 5, 7, 17, 50, 51, 73, 74, 76, 81, 97, 110, 133, 143, 147, 149
supply chain, 147, 151
surplus, 44, 133
survival, 18, 54
sustainability, 82, 111, 144
Switzerland, 94, 97
symmetry, 25, 62
synthesis, 18, 21, 54, 58

T

tanks, 91
target, 3, 15, 16, 17, 18, 20, 23, 31, 32, 34, 45, 54, 55, 56, 72, 78, 90, 91, 92, 94, 96, 97, 98, 99, 100,

102, 106, 107, 108, 110, 115, 116, 117, 119, 120, 121, 122, 123, 141, 142, 143, 144, 145, 147, 156, 171, 173
teams, 72, 128
technetium, 16, 50, 51, 76, 94, 104
techniques, 6, 18, 22, 25, 36, 44, 45, 48, 54, 59, 62, 64, 69, 70, 74, 76, 82, 85, 99, 100, 101, 115, 119, 120, 122, 123, 126, 127, 131, 132, 133, 142, 147, 150, 156, 163
technologies, 3, 9, 17, 22, 29, 31, 33, 52, 56, 59, 65, 73, 85, 99, 106, 118, 120, 123, 132, 173
technology, vii, 1, 3, 8, 9, 25, 27, 29, 32, 45, 63, 65, 70, 74, 76, 80, 84, 85, 94, 96, 99, 118, 120, 122, 126, 128, 129, 131, 156, 173
technology transfer, 94
temperature, 25, 26, 28, 30, 63, 120, 172
temporal window, 23, 171
terrestrial ecosystems, 21, 58
terrorism, 125, 128
testing, 2, 16, 18, 29, 30, 32, 54, 69, 70, 90, 112, 150, 172
thallium, 78
therapeutic agents, 23, 171
therapy, 1, 5, 12, 16, 18, 20, 23, 43, 55, 56, 57, 73, 75, 94, 106, 118, 122, 124, 125, 142, 145, 148, 154, 171
thermodynamic equilibrium, 31, 71
thorium, 2, 23, 32, 60, 173
thyroid, 18, 50, 55, 78, 102
thyroid cancer, 18, 50, 55
tissue, 2, 18, 19, 20, 55, 56, 57, 71
toxicity, 20, 56
trace elements, 34
trade, 44, 49
trafficking, 23, 171
trainees, 125
training, 4, 6, 9, 33, 46, 50, 75, 76, 96, 102, 103, 112, 116, 124, 125, 126, 130, 131, 132, 140, 146, 153, 157, 174
training programs, 125, 130
transformation, 61
transformations, 24
transition metal, 68
transmission, 70
transparency, 70
transport, 30, 45, 71, 76, 106, 108, 145, 150, 151, 158
transportation, 16, 46, 89, 92, 94, 97, 101, 108, 109, 112, 147, 151, 152, 153, 157
treatment, 8, 16, 44, 56, 59, 77, 84, 89, 94, 105, 106, 115, 134, 142, 147, 150
trial, 120, 148
tumor, 18, 19, 20, 54, 55, 56, 57, 125

U

tumor cells, 56
tumors, 18, 19, 20, 54, 55, 57, 120

U.S. economy, vii
U.S. Geological Survey, 10
Union Carbide, 133
unit cost, 147
united, 5, 6, 7, 8, 9, 22, 29, 36, 48, 51, 52, 59, 69, 74, 81, 103, 104, 106, 110, 111, 115, 117, 122, 125, 127, 159, 163
United Nations, 5
United States, 6, 7, 8, 9, 22, 29, 36, 48, 51, 52, 59, 69, 74, 81, 103, 104, 106, 110, 111, 115, 117, 122, 125, 127, 159, 163
universities, 9, 13, 31, 89, 126, 127, 131, 140, 149, 153
up-front costs, 75
uranium, 23, 45, 53, 60, 61, 69, 70, 74, 76, 80, 112, 122, 133, 158
US economy, vii, 1
USA, 27, 65, 75, 125

V

vacancies, 27, 65
vacuum, 85, 141
vapor, 91, 119, 146
variations, 23, 26, 27, 60, 63, 64
vascular surgery, 106
vehicles, 20, 56, 106
vein, 18, 54
ventilation, 78, 91, 146
Viking, 28, 65, 68
visualization, 125, 127
vulnerability, 3, 32, 45, 76, 158, 173

W

war, 133
Washington, 10, 17, 89, 101, 119, 120, 127, 131
waste, 16, 17, 30, 61, 69, 73, 75, 76, 90, 91, 92, 93, 96, 99, 107, 117, 145
waste disposal, 16, 73, 90, 117, 145
water, 23, 26, 60, 63, 67, 68, 78, 81, 91, 96, 98, 112
weapons, 8, 9, 30, 31, 45, 53, 67, 70, 71, 72, 74, 84, 85, 93, 116
weapons of mass destruction, 9, 85
web, viii, 5, 44, 46, 48, 49, 149, 153, 157
welding, 152
windows, 91, 140
Wisconsin, 10, 17, 125, 127
withdrawal, 50, 75
workers, 129
workforce, 2, 3, 33, 45, 46, 76, 102, 112, 124, 125, 126, 127, 128, 129, 132, 140, 146, 153, 154, 155, 156, 157, 174
working groups, 148
worldwide, 11, 28, 76, 78, 143, 172

X

xenon, 7, 8, 83, 84
x-rays, 70, 74

Y

Yale University, 131
yield, 1, 18, 21, 43, 54, 58, 117, 119, 145
yttrium, 56